Managing Forest Ecosystems

Volume 42

Series Editors
Margarida Tomé, Instituto Superior de Agronomía, Lisboa, Portugal
Thomas Seifert, Faculty of Environment and Natural Resources,
University of Freiburg, Freiburg, Germany
Mikko Kurttila, Natural Resources Institute, Helsinki, Finland

The aim of the book series Managing Forest Ecosystems is to present state-of-the-art research results relating to the practice of forest management. Contributions are solicited from prominent authors. Each reference book, monograph or proceedings volume will be focused to deal with a specific context. Typical issues of the series are: resource assessment techniques, evaluating sustainability for even-aged and uneven-aged forests, multi-objective management, predicting forest development, optimizing forest management, biodiversity management and monitoring, risk assessment and economic analysis.

Lauri Hetemäki • Jyrki Kangas • Heli Peltola
Editors

Forest Bioeconomy and Climate Change

Editors
Lauri Hetemäki
European Forest Institute
Joensuu, Finland

Faculty of Agriculture and Forestry
University of Helsinki
Helsinki, Finland

Heli Peltola
School of Forest Sciences
University of Eastern Finland
Joensuu, Finland

Jyrki Kangas
School of Forest Sciences
University of Eastern Finland
Joensuu, Finland

ISSN 1568-1319   ISSN 2352-3956   (electronic)
Managing Forest Ecosystems
ISBN 978-3-030-99208-8   ISBN 978-3-030-99206-4   (eBook)
https://doi.org/10.1007/978-3-030-99206-4

This Springer imprint is published by the registered company Springer Nature Switzerland AG
The registered company address is: Gewerbestrasse 11, 6330 Cham, Switzerland

Foreword

In the past 50 years, the human population has doubled and the global economy has grown nearly fourfold, together boosting the demand for food, energy and materials. The biosphere, upon which humanity depends, has been altered to an unparalleled degree.

The COVID-19 pandemic has been yet another wake-up call for us to stop exceeding the planetary boundaries. After all, deforestation and biodiversity loss have been identified as key processes in enabling the direct transmission of zoonotic infectious diseases. However, it is important to highlight that such a pandemic is not yet another global crisis, but rather one more consequence—like biodiversity loss or climate change—of the same fundamental problem—our economic system. This system is addicted to fossil resources and growth at all costs, and has failed to value our most important capital—nature.

Having arrived at the present tipping point, we clearly need a new economic paradigm that puts the basis for human prosperity within the planetary boundaries. How to transition towards this is the tricky part. Due to the speed and scale of change needed, the greatest economic transformation in human history is required in order to achieve a climate-neutral, circular and inclusive economy that prospers in harmony with nature. This paradigm shift requires *transformative policies, mission-oriented innovation and investments* in bio-based solutions and natural capital, while business models and markets need to be rethought, as well as production and consumption cycles. Above all, we need to address the past failure of our economy to value nature because our health and well-being fundamentally depend on it.

The circular bioeconomy paradigm is the new paradigm that we need. It builds on the synergies of the circular economy and bioeconomic concepts—two concepts that have so far been developed in parallel, but that now need to be connected in order to transform our economy. A circular bioeconomy offers a conceptual framework for enhancing and sustainably managing our renewable natural capital to holistically transform our land, food and industrial systems, in addition to reimagining our cities and creating new jobs and prosperity.

In such a new economic paradigm, our forests are called on to play a new and catalytic role because they are:

- The main hosts for terrestrial biodiversity
- The largest terrestrial carbon sinks
- The main terrestrial source of precipitation
- The largest source for non-food and non-feed biological resources

However, unlocking their potential requires a new vision that sees our forests not as a tool to 'compensate' for the existing broken economic system but rather as a transformative system for inspiring and creating the new economic paradigm that we need, one where life, and not consumption, becomes its true engine and its true purpose.

This book is the most crucial science-based milestone in that direction, as it provides a holistic understanding of why, what and how forests, forestry and forest-based solutions can contribute to the development of a climate- neutral, circular-bioeconomy paradigm. I would like to congratulate all the authors of the book for putting together this valuable work. Moreover, I want to thank those authors, such as the Assistant Director of the EFI, Lauri Hetemäki, who have contributed to creating an honest, open and informed dialogue between different stakeholders and scientific disciplines within and between countries (and in the EU) in order to realise the full potential of forest-related science for sustainable action.

European Forest Institute (EFI)   Marc Palahí
Joensuu, Finland

Aim and Scope

Climate change, global population growth, declining natural resources and the loss of biodiversity challenge us to move towards a global bioeconomy, based on the sustainable utilisation of renewable natural resources in the production of energy, products and services. The linear economic model based on fossil raw materials and products is coming to an end. Major global agreements and policy goals—the Paris Climate Agreement and the United Nations Sustainable Development Goals—have given licence for our economic model to be changed. There is the need for a new economic paradigm that will place the basis for human prosperity within the planetary boundaries. One essential part of this new paradigm has to be a forest-based circular bioeconomy.

The shift to this bio-based economic paradigm should be a long-term strategy for decoupling economic growth from climate change and environmental degradation. Developments in science and technology are laying the foundations for the bioeconomic age. Bio-based products have already emerged that can substitute for fossil-based materials, such as plastics, chemicals, textiles, cement and many other materials. Now, the big question is how to turn these scientific and technological successes into a global economic paradigm shift, and in a sustainable way. This requires us to look at the potential synergies and trade-offs that such a change will inevitably bring and how these can be integrated with the economic, ecological and social goals of society.

Right now, we know that climate change will take place in this century, although there is uncertainty as to the degree of disruption it will bring. It will have an impact on forests. Like humans, trees are mortal. Climate change threatens to increase the mortality rate of trees. Disturbances, such as droughts, fires, storms and bark- beetle outbreaks, have already become stronger, more extensive and more damaging. This trend requires us to adapt to climate change and to build resilience in our forests against climate change. So, how can we do this?

These themes and questions are the focus of this book, which builds upon recent scientific evidence concerning forests and climate change, and examines how the development of a forest bioeconomy can help to address the grand challenges of our time. In the book, experts analyse the economic, ecological and social dimensions of forests and climate change, along with the basis for, and shaping of, a forest-based bioeconomy, and the links between these. In this way, it provides information on the potential of forests and forest-based products to help in mitigating climate change, and the types of measures that can be taken to adapt forests to climate change, thereby building forest resilience. The book outlines a *climate-smart forestry* approach, based on three main objectives. First, reducing net emissions of greenhouse gases into the atmosphere. Second, adapting and building forest resilience to climate change. Third, sustainably increasing forest productivity and economic welfare based on forestry. The climate-smart forestry approach is illustrated by case studies from Czech Republic, Finland, Germany and Spain—countries that have quite different forests and forest sectors. Finally, we suggest the types of policy measures required to address the challenges of developing, and increase the opportunities associated with, a sustainable forest bioeconomy.

To the best of our understanding, this is the first book devoted to examining the links between climate change and a forest bioeconomy, and outlining the need for a climate-smart forestry approach to address the many needs we have for forests. The book is directed at forest- and environment-sector stakeholders and decision- makers, as well as the research community, the broader education sector and the media.

Acknowledgements

This book is based mainly on work stemming from the FORBIO project 'Sustainable, climate-neutral and resource-efficient forest-based bioeconomy', which was funded by the Strategic Research Council (SRC) of the Academy of Finland. Support for finalising the book was provided under Finland's UNITE Flagship 'Forest– Human– Machine Interplay—Building Resilience, Redefining Value Networks and Enabling Meaningful Experiences'. Funding from the Academy of Finland (grant number 314224 for SRC FORBIO project 2015–2021, and grant number 337127 for UNITE Flagship 2020–2024) is also gratefully acknowledged. The Finnish Forest Foundation is thanked for enabling the open-access publication of this work. Finally, we would like to express our gratitude to Janni Kunttu from the European Forest Institute for helping us with the administration and technical editing of the book.

Contents

Editors and Contributors

About the Editors

Lauri Hetemäki is Professor of Practice at the Faculty of Agriculture and Forestry, University of Helsinki, and a senior researcher at the Circular Bioeconomy Alliance, European Forest Institute (EFI). He has a PhD in economics from the University of Helsinki. Previously, he has worked as the assistant director of EFI, Professor of Forest Sector Foresight at the University of Eastern Finland, and as a senior scientist at the Finnish Forest Research Institute. He has extensive experience in science-policy work, e.g., in supporting the work of the European Parliament, the European Commission, and the Finnish Government and Parliament, as well as coordinating pan-European science-policy studies. He is an author of over 250 scientific and popular publications. Hetemäki is a Fellow of the Royal Swedish Academy of Agriculture and Forestry.

Jyrki Kangas is Professor of Forest Bioeconomy at the University of Eastern Finland and director of the Forest-Human-Machine Interplay (UNITE) Flagship of Science. He has a PhD in forest sciences from the University of Joensuu. He was the director general of Metsähallitus (2006–2014), director of forestry at UPM (2003–2006) and CEO of UPM's real estate daughter company Bonvesta (2005–2006), and Professor of Forest Planning and a research station head at Finnish Forest Research Institute (1994–2003). He is an author of over 300 scientific and popular publications. Kangas has been awarded the Scientific Achievement Award by IUFRO. His positions of trust have included vice president of EUSTAFOR, chair of Koli Forum, and chair of the Foundation for European Forest Research, among others. Currently, he is a member of the board of directors of Business Joensuu Ltd and of A. Ahlström Real Estate Ltd. Kangas was FORBIO's vice PI and interaction coordinator.

Heli Peltola is Professor of Silvicultural Sciences at the Faculty of Science and Forestry, University of Eastern Finland. She has a PhD in silvicultural sciences from the University of Joensuu. Previously, she was a member of the Research Council for Biosciences Health and the Environment (BTY 2019–2021). Currently, she is a member of the Finnish Climate Change Panel (2020–2023). She has extensive experience in leadership of research projects, including the SRC FORBIO project from 2015 to 2021. She is an author of over 200 scientific peer-reviewed publications. Peltola was awarded with the decoration Knight First Class of the Order of the White Rose of Finland by the President of the Republic of Finland in 2020, and with the Carola and Carl-Olof Ternryd's award by the Linnaeus Academy's Research Foundation in 2020.

Contributors

Aitor Ameztegui The Forest Science and Technology Centre of Catalonia (CTFC), Solsona, Spain University of Lleida, Lleida, Spain

Perttu Anttila Natural Resources Institute Finland (Luke), Helsinki, Finland

Antti Asikainen Natural Resources Institute Finland, Joensuu, Finland

Emil Cienciala IFER – Institute of Forest Ecosystem Research, Jílové u Prahy and Global Change Research Institute of the Czech Academy of Sciences, Brno, Czech Republic

Jose Ramón González The Forest Science and Technology Centre of Catalonia (CTFC), Solsona, Spain

Elena Górriz-Mifsud The Forest Science and Technology Centre of Catalonia (CTFC), Solsona, Spain

Marc Hanewinkel University of Freiburg, Freiburg, Germany

Tero Heinonen University of Eastern Finland, Joensuu, Finland

Lauri Hetemäki European Forest Institute, Joensuu, Finland
Faculty of Agriculture and Forestry, University of Helsinki, Helsinki, Finland

Elias Hurmekoski University of Helsinki, Helsinki, Finland

Janne Jänis University of Eastern Finland, Joensuu, Finland

Jyrki Kangas University of Eastern Finland, Joensuu, Finland

Sari Karvinen University of Eastern Finland, Joensuu, Finland

Antti Kilpeläinen University of Eastern Finland, Joensuu, Finland

Janni Kunttu European Forest Institute, Joensuu, Finland

Mikko Laapas Finnish Meteorological Institute, Helsinki, Finland

Ilari Lehtonen Finnish Meteorological Institute, Helsinki, Finland

Andrey Lessa Derci Augustynczik University of Freiburg, Freiburg, Germany

Antti Mutanen Natural Resources Institute Finland, Helsinki, Finland

Heli Peltola University of Eastern Finland, Joensuu, Finland

Kimmo Ruosteenoja Finnish Meteorological Institute, Helsinki, Finland

Jyri Seppälä Finnish Environment Institute, Helsinki, Finland

Olli-Pekka Tikkanen University of Eastern Finland, Joensuu, Finland

Antoni Trasobares The Forest Science and Technology Centre of Catalonia (CTFC), Solsona, Spain

Ari Venäläinen Finnish Meteorological Institute, Helsinki, Finland

Hans Verkerk European Forest Institute, Joensuu, Finland

Jari Viitanen Natural Resources Institute Finland, Helsinki, Finland

Rasoul Yousefpour University of Freiburg, Freiburg, Germany

Prologue

Isn't it surprising that dead forests – for coal results from the decomposition of forests – should annually supply a larger volume of raw material than live forests?

The above statement, by Egon Glesinger some 70 years ago in his visionary book, *The Coming Age of Wood*, is still true today (Glesinger 1949, p. 21).[1] In 2019, global coal production was around twice that of roundwood production.[2] This at a time when, for several decades, we had already understood that coal was a major driver of climate change, which was *not* common knowledge at the time of Glesinger's writing. His book reflected the situation following World War II—the need to rebuild much of the global infrastructure and improve the welfare of war-ravaged people. Glesinger was influenced by Germany's innovative efforts to build an economy based on wood. Germany's motivation was to enhance self-sufficiency in transportation fuels, chemicals, feedstock and other critical raw materials and products at a time when there were increasing risks to their supply. In a pioneering way, Germany advanced and utilised wood chemistry for these purposes.

Glesinger saw the role of wood in a much more extensive and diverse way than was generally the case 70 years ago, or may sometimes be the case even today. He detailed three major reasons for using more wood and why it was unique among all raw materials—wood is universal (it serves many requirements of human existence), wood is abundant (forests cover a major proportion of Earth's land area) and wood is inexhaustible (with the proper management of forests, wood is renewable). Thus, unlike coal, natural gas or oil, forests (wood) are not mines that will eventually be depleted.

---

[1] Glesinger E (1949) The Coming Age of Wood. Simon and Schuster, Inc., New York. http://www.archive.org/details/comingageofwood00gles. Accessed 20 Jan 2021.

[2] In 2019, global coal production amounted to 7921 Mt (International Energy Agency 2020) and roundwood production 3964 Mm^3 (FAOSTAT), which, according to a rough estimate, was around 3500 Mt. Converting roundwood cubic metres to tonnes can only be an approximation due to the weight of wood varying across tree species and timber type.

Today and tomorrow, the need to feed, clothe, package, transport and build—that is, secure the basic necessities of life—will remain. For example, it has been projected (UN 2019) that, by the end of this century, there will be 3 billion more people to be fed and more than 2 billion new homes to be built (Smith 2018). Glesinger argued that wood should be viewed as central to the satisfaction of human needs. This is even more true today, since we want to also satisfy these needs with *sustainable production* and *consumption*, which we have not been able to do in the past 70 years. A large part of sustainability is to do it in a way that does not affect our climate, but rather helps to mitigate the ongoing climate change.

In this book, we argue that to marry human needs with sustainability is not possible without using also biological resources for those needs. In fact, we posit that a circular bioeconomy is an essential tool—even if insufficient—for facilitating the movement to sustainable development and reaching the goals world states have set for climate-change mitigation and societal welfare.

Lauri Hetemäki

Chapter 1
Forest Bioeconomy, Climate Change and Managing the Change

Lauri Hetemäki and Jyrki Kangas

Abstract In order to realise Agenda 2030, or the United Nations' Sustainable Development Goals, and the Paris Climate Agreement, the business-as-usual model—the policies, production and consumption habits we have been following thus far—will not work. Instead, it is necessary to change the existing economic model and how we advance societal well-being. Here, we argue that a forest-based bioeconomy will be a necessary, albeit insufficient, part of this transformation. The European forest-based sector has significant potential to help in mitigating climate change. However, there is no single way to do this. The means to accomplish this are diverse, and these measures also need to be tailored to regional settings. Moreover, the climate mitigation measures should be advanced in synergy with the other societal goals, such as economic and social sustainability. Climate mitigation in the forest- based sector requires a holistic perspective.

Keywords Circular bioeconomy · Transformation · Forests · Climate change mitigation · Synergies · Trade-offs

1.1 Introduction

The world states agreed, in 2015, on Agenda 2030, or the United Nations' (UN) Sustainable Development Goals (SDGs), and the Paris Climate Agreement. It is widely agreed that the business-as-usual model—the policies, production and

L. Hetemäki (✉)
European Forest Institute, Joensuu, Finland

Faculty of Agriculture and Forestry, University of Helsinki, Helsinki, Finland
e-mail: lauri.a.hetemaki@helsinki.fi

J. Kangas
University of Eastern Finland, Joensuu, Finland

© The Author(s) 2022
L. Hetemäki et al. (eds.), *Forest Bioeconomy and Climate Change*, Managing
Forest Ecosystems 42, https://doi.org/10.1007/978-3-030-99206-4_1

consumption habits we have been following thus far—will not help us to reach these goals. These agreements and goals can therefore be interpreted as providing a mandate to change the existing economic model–how we advance societal well-being. In this book, *we argue that a forest-based bioeconomy is a necessary part of this transformation.*

There are many definitions of the bioeconomy, as well as usage of similar terms, such as *biobased economy* and *green economy* (D'Amato et al. 2017). In practice, the bioeconomy has turned out to be a changing concept and adjustable for various purposes. One useful definition is from the Global Bioeconomy Summit (GBS) 2015: "*bioeconomy as the knowledge-based production and utilization of biological resources, innovative biological processes and principles to sustainably provide goods and services across all economic sectors*". The bioeconomy therefore encompasses the traditional bioeconomy sectors, such as forestry, paper and wood products, as well as emerging new industries, such as textiles, chemicals, new packaging and building products, biopharma, and also the services related to those products (research and development, education, sales, marketing, extension, consulting, corporate governance, etc.), and forest services (recreation, hunting, tourism, carbon storage, biodiversity, etc.).

Hetemäki et al. (2017) extended this definition to a *circular bioeconomy*, also linking it to the *natural-capital* concept. A circular bioeconomy builds on the mutual efforts of the circular economy and bioeconomy concepts, which in many ways are interlinked. The European Environment Agency (EEA) has indicated that implementing the concepts of a bioeconomy and circular economy together as a systemic joint approach would improve resource efficiency and help reduce environmental pressures (EEA 2018). We further suggest that these two concepts, which are often considered separately, could create marked synergies when applied as a hybrid approach, making simultaneous use of both, as is the concept of the circular bioeconomy.

In this book, we understand bioeconomics along similar lines to the GBS (2015), and the extension of this introduced by Hetemäki et al. (2017). We particularly emphasise three key aspects of bioeconomics:

- the transformational role of the bioeconomy in helping to mitigate climate change, and to replace fossil-based products (e.g. oil-based plastics and textiles), non-renewable materials (e.g. steel, concrete) and non- sustainable biological products (e.g. cotton in certain regions);
- the enhancement of the *natural-capital* approach to the economy, involving better integration of the value of natural resources and life-sustaining regulatory systems (e.g. biodiversity, freshwater supplies, flood control) with economic development, as suggested by Helm (2015) and in the action plan of Palahí et al. (2020); and
- the improvement of the quality of economic growth, making it sustainable and operating in synergy with SDGs rather than trade-offs.

The first aspect is generally already well understood in bioeconomic strategies, the latter less so. The long-term sustainable production of natural capital relies on the key role of forests as the most important land-based biological infrastructure on the

European continent (see Chap. 1, Box 1.1). Forests provide the largest supply of renewable biological resources not competing with food production (unlike biomass from agricultural land). Moreover, combining digital technology with biology can offer increasing opportunities for the bioeconomy in the future.

Although the concepts used in the chapter title—*bioeconomy* and *climate change*—have attendant ambiguities, and there is a scientific discourse concerning what they actually mean (e.g. Hulme 2009; Kleinschmit et al. 2017), these terms are not discussed here. Rather, given the above definition of the bioeconomy, we examine its substance in the context of the forest-based sector, examining how it can be implemented, what the outlook is, and its relationship with climate change. In turn, we perceive climate change as global warming and its effects. We also follow the understanding of the Intergovernmental Panel on Climate Change (IPCC), which has posited a human influence on climate that has been the dominant cause of the observed warming since the mid-twentieth century.

Moreover, this book highlights the forests of the European Union, although many of the issues and implications discussed could probably be generalised to other regions. Yet, when discussing climate change and forests, it is important to acknowledge some key distinctions between world regions. For example, there are major differences between tropical forests (45% of the world total in 2020), boreal forests (27%), temperate forests (16%) and subtropical forests (11%) (Food and Agriculture Organization [FAO] 2020), and between the institutional settings in which these forests are located. For example, in Europe, the forests are mainly boreal and temperate, whereas in South America, they are tropical. Moreover, in terms of the environmental opportunities and challenges that climate change is promoting, and the institutional contexts of the continents, contrasting measures may need to be prioritised more in South America than in Europe. The crudest and simplest way to illustrate this point is to look at the forest statistics from the last three decades (see Table 1.1).

Table 1.1 Different trends in forest development in Europe and South America, 1990–2020

Variable (unit)	1990	2000	2010	2020
European forests 1990–2020 *(50 countries and territories)*				
Forest area (million ha)	994	1002	1014	1017
Forest area (% of land area)	44.9%	45.3%	45.8%	46.0%
Growing stock (billion m³)	104	108	113	116
Carbon stock in biomass (Gt)	45	48	51	55
Total carbon stock (Gt)	159	162	168	172
South American forests 1990–2020 *(14 countries and territories)*				
Forest area (million ha)	974	923	870	844
Forest area (% of land area)	55.8%	52.8%	49.8%	48.3%
Growing stock (billion m³)	207	199	191	187
Carbon stock in biomass (Gt)	106	102	98	96
Total carbon stock (Gt)	162	155	148	145

Data Source: FAO (2020)

In terms of forest area (ha), Europe and South America were almost of equal size in 1990 – Europe's forest area was only 2% larger. From 1990 to 2020, the European forest area grew by 2.3% and the carbon stock in the forest biomass by 18% (Table 1.1). However, exactly the opposite trend took place in South America, where the forest area and carbon stock have declined by 13.3% and 3.3%, respectively (Table 1.1).[1] Thus, today, the European forest area is one-fifth bigger than that of South America. One clear implication from these statistics is that South America should focus on reversing its deforestation trend in order to better contribute to climate-change mitigation (among other things), whereas in Europe, the priority might not be so much the forest area, but rather other mitigation measures, which this book will discuss in more detail.

When discussing forests and climate change, sometimes the media, and even some scientists, seem to forget these differences in opportunities and challenges that distinct forests and continents are facing deforestation or declining carbon stock may not to be the priority issue in European boreal and temperate forests. Moreover, forests and climate change together present complex issues, and there appears to be no silver bullet that would work in all circumstances and regions, even within Europe (Hulme 2009; Nabuurs et al. 2017; Nikolakis and Innes 2020).

The diversity, complexity and feedback effects among the different channels through which forest-based-sector mitigation can be increased have not always been well understood in the discussion. Rather the media reporting, and occasionally the scientists' messages to policy-makers, have tended to narrow and simplify the topic in a way that misses the holistic picture (Hetemäki 2019; Chapters 8 and 9). For example, the links between climate mitigation and adaptation, the role of forest disturbances, the socioeconomic context (techno-system) in which we are operating, the importance of considering both the short- and long-term impacts, and the need to consider climate mitigation simultaneously with the other grand challenges of humanity. The different roles of forests in climate mitigation are summarised in Table 1.2, which gives a simplified taxonomy that lists some of the most important features between the forest-based sector and climate mitigation.

In Table 1.2, any one of the channels through which the forest-based sector can impact climate mitigation points to a specific action to maximise the mitigation potential under that specific option. Thus, if for example, one was only concerned about maximising the sequestration of carbon in forests and soils, it would make sense to conserve forests, allowing no commercial harvests, at least in the short term (i.e. the coming decades). On the other hand, if the substitution impact was being emphasised, the remedy would be to increase wood production. Furthermore, if the vulnerability of forests to disturbances and damage is also considered, the complete conservation of forests and refraining from harvesting would not be recommended, especially in the long term, as forest ageing increases the probability of both abiotic damage and a number of biotic injuries to trees (see Chap. 3).

---

[1] In Africa, forest area (ha) has declined from 1990 to 2002 by 14.3%, whilst in North and Central America and Oceania, it has stayed basically the same, and in Asia, it has grown by 6.5% (FAO 2020).

Table 1.2 Forest-based-sector climate-mitigation impacts and actions to strengthen these

Mitigation channel	Possible actions to increase mitigation by 2050 *(action could be modified if the target was long term, e.g. beyond 2050)*
Forest biophysical impacts	
Forest carbon sequestration in trees and soils *(forest sink)*	Stopping deforestation, increasing afforestation and forest conservation. Turning global forest loss to forest gain, and reforestation always after final felling. Increasing tree growth, and reduce harvests.
Forest albedo	Changing coniferous forests to broadleaves or mixed forests
Forest aerosols	Afforestation and conserving forests
Forest disturbances	Adapting forests to changing climate and increasing resilience (e.g. changing tree species and provinces). Decreasing disturbance risks via forest management measures (e.g. increasing mixed forests and decreasing monocultures)
Substitution and storage impacts	
Substituting forest biomass for fossil raw materials, energy and products	Forest management and wood production for forest-based products. Policies to enhance demand for forest-based products, such as wood construction
Storing carbon in forest products	Forest management and wood production for forest-based products
Emissions from forest products value-chain	
Production and logistics	Reducing and eliminating the use of fossil fuels in transport, heating and electricity generation in forest-based industries
Socioeconomic and political impacts *(feedback impacts)*	
Synergies or trade-offs between the mitigation channel and other societal objectives *(e.g. leakage impacts, political support for mitigation measures, biodiversity impacts, income and employment impacts)*	Seek to maximise synergies and minimise trade-offs between mitigation measures and other societal goals
Combination of several different channels	No single policy/action can enhance all the different mitigation channels > need a mixture of different policy and management actions

Clearly, if all the different channels and socioeconomic and political responses are considered simultaneously and holistically, the action may be different than for any single option alone. The planning of mitigation actions is even more complicated by the fact that, depending on the time span of the policy target, different actions may be favoured. That is, if the target is short term (up to 2050) or long term (beyond 2050), the actions required might be somewhat different. Indeed, the occasionally different messages received from scientists on the most appropriate measures to mitigate climate change via forests may reflect them focusing on different time spans. Moreover, one can come to well-founded but different conclusions, depending on whether an analysis is based on looking at only one (or some) of the many possible mitigation channels, or if it is based on a holistic approach, seeking to synthesise the different impact channels and feedback loops. In this book, we follow the IPCC (2019) understanding, where the forest-based sector can contribute to climate mitigation by enhancing forest carbon stocks and sinks, storing carbon in harvested wood products, and substituting for emissions-intensive materials and fossil energy.

In general, this book is based on an approach that stresses the importance of taking a *holistic approach* to assessing how to best utilise forests to mitigate climate change. The *Climate Smart Forestry* (CSF) approach has been introduced as a means of integrating the holistic approach to increase climate mitigation via forests and the forest sector (Nabuurs et al. 2015, 2017; Kauppi et al. 2018; Yousefpour et al. 2018). It is based on acknowledging the diversity and complexity of the issue, as outlined in Table 1.2. The CSF approach seeks to connect forests to bioeconomics, link mitigation and adaption measures, enhance the resilience of forest resources and ecosystem services, while at the same time, seek to meet the other societal challenges (employment, income, biodiversity, etc.). CSF has been introduced in the European context (*see references cited above*), but the approach is of global relevance. CSF builds on the concepts of sustainable forest management, with a strong focus on climate and ecosystem services. It builds on three mutually reinforcing components:

- increasing carbon storage in forests in conjunction with other ecosystem services;
- enhancing health and resilience through adaptive forest management; and
- using wood resources sustainably to substitute for non-renewable, carbon-intensive materials.

CSF aims to incorporate a mix of these measures by developing spatially diverse forest management strategies that acknowledge all carbon pools simultaneously to provide longer-term and greater mitigation benefits, while supporting other ecosystem services. Such strategies should combine measures to maintain or increase carbon stocks in forest ecosystems and wood products, and maximise substitution benefits, while taking regional conditions into account.

1.2   What Are the Future Challenges and Opportunities?

The fact that humanity and forests are not facing only one challenge at a time, but several simultaneous environmental, societal and economic problems, points to there being no simple answers. Just think, for example, about the need to increase climate mitigation efforts, biodiversity, and employment and income opportunities for the growing population and middle class. Moreover, depending on the country or region and its particular circumstances, the needs may have a somewhat different emphasis, and the opportunities to fulfil these may also be different. Therefore, it would be unrealistic to assume that there can be a simple answer to all the needs and local opportunities. This situation is also reflected in the CSF approach, which should be tailored to local conditions (Nabuurs et al. 2017). The optimal measures taken under the CSF approach in forests and the forest- based sector can vary even among different parts of a country. This book seeks to clarify the different options under CSF, and why some measures might be preferred to others, depending on the regional specificities (Chap. 10).

It is also evident that there can be synergies and trade-offs between the many ecosystem services forest generate, or between the environmental, economic and social objectives that society demands from forests. We argue that the objective of bioeconomy strategies and policies should be to maximise the potential synergies and minimise the trade-offs between the bioeconomy, biodiversity and climate miti-gation. Hetemäki et al. (2017) illustrated the role of synergies and trade-offs (see Fig. 1.1; see also Biber et al. 2020; Krumm et al. 2020).

In economic terms, the green curves show a forest bioeconomic *production-possibility frontier*, when there is a trade-off between the outputs that forests can provide. The frontier describes all output combinations when outputs are produced efficiently. It is, of course, possible that society is operating inefficiently and would be located below the production possibility frontier.

The vertical axis in Fig. 1.1. describes non-product forest services (biodiversity, carbon sink, water quality, recreation, tourism, etc.), whilst the horizontal axis rep-resents forest products (pulp, sawnwood, bioenergy, etc.). The Fig. 1.1 illustrates a bioeconomy that can use forest resources to produce both material forest products and non- product services at the same time, and can choose between alternative combinations of each production type. The green curves—the so-called

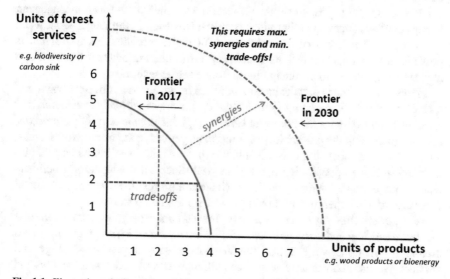

Fig. 1.1 Illustration of a forest-based, bioeconomic production-possibility frontier, with trade-offs and synergies between forest products and non-product forest services

production-possibility frontiers—indicate the maximal combination of outputs (e.g. biodiversity and pulp) for a given amount of inputs (forest, capital, labour). The location of the frontier is determined by technological constraints and resource availability. By picking any point on the green line, the respective amounts of forest products and non-product services can be read from the axes.

As Fig. 1.1 suggests, the more intensively forests are used for forest products, the less societies can produce services such as biodiversity, and vice versa. The challenge for society is to find a sustainable combination of both. The role of synergies is important, because they can move the frontier outwards, and in this way alleviate the trade-offs. In Fig. 1.1, the frontier may move outwards in two ways—either via more from more, or more from less. In both cases, more forest products and non-product services are produced. This results in more sustainable forestry, irrespective of whether the society values more forest products or non-product forest services, *ceteris paribus*. The outward movement of the frontier is, in principle, possible via three pathways:

- technological change (innovation) and learning-by-doing (e.g. better management experience);
- increased resource efficiency with given production inputs (e.g. more forest growth, capital and/or labour productivity); and
- a combination of these two.

Evidence seems to support that outward movement of the frontier is possible. For example, Bieber et al.'s (2020) European case studies indicated a considerable range of forest management options that would not automatically cause trade-offs between wood production, biodiversity and carbon sequestration, also showing options for building synergies between these. However, the new production-possibility frontier would usually not be possible in the short term, especially with forest management taking time to implement and produce changes. However, in the longer term, when technology and innovations are introduced, or higher productivity, movement is possible. Innovations and technological progress (including better institutions and management) are key to producing more from existing resources.

Figure 1.1 illustrates that the bioeconomy can be advanced in different ways, and therefore *it would be optimal to provide policy incentives that help to minimise the trade-offs and maximise the synergies between different components of the bioeconomy.* By increasing the profitability of forest management, and possibly forest areas, a well-promoted bioeconomy could enhance the possibilities of taking care of biodiversity. But the opposite is important as well. Successful adaptation to climate change and extreme weather conditions (increasing forest fires, storms, pests and other hazards) is imperative to provide a basis for the bioeconomy.

So, a key question for bioeconomics is, how can the synergies be made stronger and trade-offs reduced using policies and different measures in forests and the forest sector? In this book, we examine this question, seeking to provide some general answers. We also show that this may mean different actions in disparate regions and under contrasting circumstances.

Finally, the above approach also requires the need to abolish the conventional and still-dominant thinking in which the economy (e.g. wood production) and the environment (e.g. biodiversity) are seen as necessarily and fundamentally opposed to each other. Certainly, there are plenty of cases in the past in which this has been true, as it can be in the future. However, it would be much more fruitful to start to find ways to embrace the synergies than could exist between the economy and the environment. In this book, we argue that the circular bioeconomy, and more specifically CSF, can be an approach for enhancing these synergies and minimising trade-offs in the forest- based sector.

1.3 Outline of the Book

To the best of our knowledge, this is the first comprehensive book to examine the forest bioeconomy and its connection to climate mitigation and adaptation in the EU forests and forest-based sector. It also describes how the CSF approach is a useful tool for combining bioeconomics and climate mitigation. The CSF approach is illustrated using countries that differ in terms of their forest sectors as case studies. The focus is on the EU context, but the principles of the approach may be tailored to other regions.

The analysis in the book is based significantly on the results of an interdisciplinary research consortium project funded by the Strategic Research Council of the Academy of Finland, *Sustainable, climate-neutral and resource- efficient forest-based bioeconomy* (FORBIO) that was carried out in 2015–2021. Needless to say, not all the wisdom on the topic presented in this book lies in the findings of one research project. To try and address this shortcoming, the analyses in the book also refer to the international scientific literature and syntheses of this, such as the IPCC assessment reports. Authors outside the FORBIO project have also contributed to the analyses.

In terms of forest and climate mitigation analysis and discussion, this book endeavours to show the complexity and diversity of the ways in which that can take place, linking these to other demands placed on forests by society. It describes the individual mitigation channels in detail in the different chapters. However, in those chapters discussing the implications of policy and forest management measures, the perspective is typically holistic. That is, for policy and forest management measures, it is necessary to consider the implications of all the individual mitigation channels at the same time, and find an optimal balance between these actions, which individually may even point to opposing measures. In summary, all the different mitigation channels shown in Table 1.2 and the societal context should be kept in mind.

Box 1.1 Forest Bioeconomy in the EU

Antti Mutanen and Jari Viitanen
Natural Resources Institute Finland, Joensuu, Finland

The EU's updated Bioeconomy Strategy promotes bioeconomy as a means of tackling global challenges, such as climate change, ecosystem degradation and the unsustainable consumption of natural resources, while simultaneously supporting the modernisation of European industries and strengthening Europe's competitiveness in global markets (European Commission [EC] 2018). The objectives of the updated Bioeconomy Strategy (ensuring wood security, managing natural resources sustainably, reducing dependency on non-renewable resources, mitigating and adapting to climate change, strengthening European competitiveness and creating jobs) are the same as in the original Bioeconomy Strategy of 2012. However, in the updated strategy, the concepts of sustainability and circularity are emphasised as being at the core of the bioeconomy, and are integral prerequisites for the acceptability and future success of the bioeconomy. In fact, while recognising the need for recycling and waste streams as an alternative source of biomass, the original Bioeconomy Strategy did not address circularity or circular economy explicitly (EC 2012).

Despite the emphasis on the ecological dimension of sustainability, the updated Bioeconomy Strategy identifies competitiveness and job creation, representing economic sustainability, as key drivers of bioeconomy. By 2030, the strategy envisages the creation of one million new jobs in bio-based industries (EC 2018). These jobs would emerge in rural and coastal areas especially.

The Joint Research Centre (JRC) publishes bioeconomy statistics for the whole EU and its individual member states. These statistics are based on the data collected in the European Statistical System (ESS). The ESS employs the NACE Rev. 2 classification in its collection of data on economic activity from different fields of the economy. The development of the NACE classification began in the 1970s, and this division of the economy into different industries and sectors is well established, reflecting the traditional way of classifying a multitude of economic activities for the needs of sectoral policymaking. However, the bioeconomy crosses the boundaries of traditional sectors, and the statistics based on the standard classification are inadequate to provide a comprehensive picture of the scale and trends of the bioeconomy. Thus, in the JRC's bioeconomy statistics, some of the NACE Rev. 2 sectors, such as agriculture, fisheries, food, forestry, wood products, and pulp and paper production, are included entirely in the bioeconomy, while only the bio-based share of other sectors, such as the chemical, biotechnological and energy industries, is included (for more details, see Ronzon et al. 2017).

(continued)

Box 1.1 (continued)

According to the JRC's statistics, the bioeconomy created €614 billion value added and employed 17.5 million people in the EU27 in 2017 (JRC DataM 2021). The bioeconomy's contribution to the gross domestic product (GDP) was, on average, 4.7% across all the EU27 member states, but the variation between the member states was substantial (Fig. Box 1.1). In Lithuania, the contribution of the bioeconomy to the national GDP was the highest, at 8.1%, whereas in Luxemburg, it was the lowest at 0.8%. Even greater variation between the member states can be detected in the bioeconomy's contribution to employment. On average, the number of employees in the bioeconomy was 8.9% of the total number of employees across all sectors in the EU27, while the share was the highest, at 27.8%, in Romania, and the lowest, at 3.5%, in Luxemburg. Geographically, the bioeconomy's role in the national economies tends to be higher than average in the Eastern

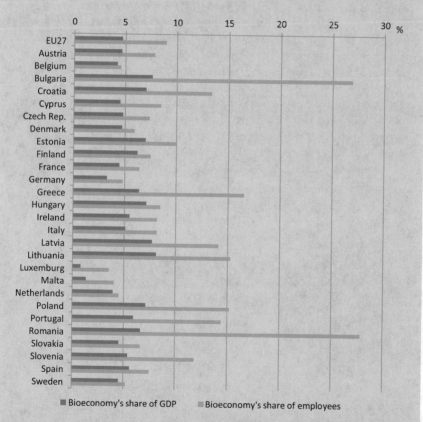

Fig. Box 1.1 The bioeconomy's share of GDP and number of employees in the EU27 in 2017. (Sources: JRC DataM 2021 and Eurostat 2021)

(continued)

Box 1.1 (continued)

European countries, where the agricultural sector is large compared to other sectors, especially in terms of people employed.

By sub-sector, the EU's bioeconomy is dominated by agriculture and the food industry. Agriculture accounted for 53% of employment and 31% of value added in the EU27 bioeconomy in 2017, while the corresponding figures for the food industry and the production of beverages and tobacco were 25 and 35% (Fig. Box 1.2). At the same time, the forest bioeconomy, including forestry, wood products, furniture, and the pulp and paper industries, generated 19% of the total value added and employed 2.5 million people – that is, 14% of the total number of employees in the EU27 bioeconomy. The importance of the forest bioeconomy varies greatly between the member states. In Austria, Finland, Sweden, Estonia, Latvia, Slovakia and Slovenia, the role of the forest bioeconomy is especially pronounced, generating more than 30% value added in the national bioeconomy (Fig. Box 1.3).

In the EU Bioeconomy Strategy, the potential of the forest bioeconomy is recognised as a source of raw materials that could replace fossil materials in the construction, packaging, furniture, textile and chemical industries. Emphasis is also placed on new business models based on the ecosystem services provided by forests, such as carbon storage and sequestration, water regulation and business opportunities in nature tourism. However, the possibility of increasing harvesting volumes, even without exceeding the annual increment, is treated with caution, since trade-offs between the use of woody biomass and other ecosystem services are considered significant and have to be analysed carefully.

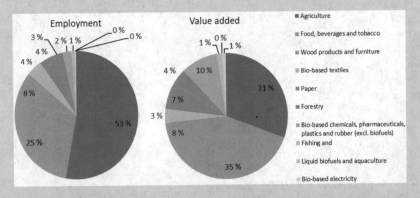

Fig. Box 1.2 Shares of employment and value added by sub-sector in the EU27 bioeconomy in 2017. (Source: JRC DataM 2021)

(continued)

Box 1.1 (continued)

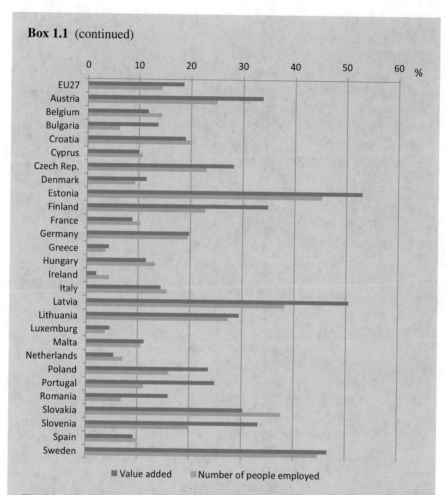

Fig. Box 1.3 The forest bioeconomy share of value added and employment in the EU27 in 2017. The forest bioeconomy covers forestry, the manufacture of wood products and wooden furniture, and of pulp and paper. (Source: JRC DataM 2021)

The definition of bioeconomy and its boundaries in relation to traditional industries and classification framework are not unambiguous, and the JRC's database is only one source of bioeconomy statistics. For example, Natural Resources Institute Finland (Luke) has recently started publishing Finnish bioeconomy statistics, according to which the share of value added created by bioeconomy was 12.2% of the Finnish GDP in 2017, a figure almost twice as high as the estimate provided by JRC (Fig. 1.1). The discrepancy between Luke's and JRC's figures is solely due to the differences in the industries and the proportions of industries included in the bioeconomy. Kuosmanen et al. (2020) studied the relevant industries to be included in the bioeconomy at EU level and proposed a method of determining the size of bioeconomy which combines both the input- and output-based approaches in contrast to the

(continued)

Box 1.1 (continued)

purely output-based approach employed by the JRC and Luke. The analysis emphasized the importance of bio-based services, i.e. tertiary sector, such as construction sector, restaurants, and transportation of bio-based goods. According to Kuosmanen et al. (2020), the value added of bioeconomy in the EU28 was EUR 1,460.6 billion or 11% of the GDP in 2015. Robert et al. (2020) studied wood-based bioeconomy in the EU28, and the results stressed the importance of secondary processing, such as wood-based construction,

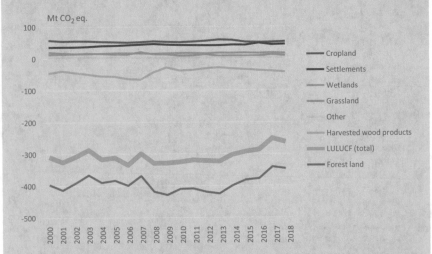

Fig. Box 1.4 GHG net emissions (+) and removals (−) from the LULUCF sector and its sub-sectors in the EU27 in 2000–2018. The 'Other' category includes the following sectors: other land; other land use; land-use change; forestry; and indirect N_2O emissions from managed soils. (Source: EEA 2021)

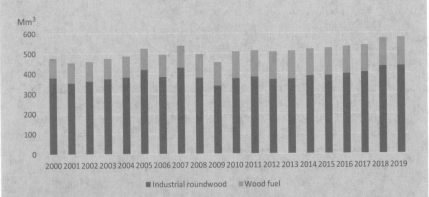

Fig. Box 1.5 Production volumes (overbark) of industrial roundwood and wood fuel in the EU27 in 2000–2019. Underbark figures were converted to overbark using the coefficient 1/0.88 (FAO, International Tropical Timber Organization [ITTO] and UN 2020). (Source: FAO 2021)

(continued)

Box 1.1 (continued)

printing, and wood-based energy production, in the creation of jobs, and the total number of people employed in the wood-based bioeconomy was 4.5 million in 2018. Obviously, there is a need for developing the statistics to provide the decision makers with comprehensive and consistent data on the scope and trends of bioeconomy.

The role of forests in combating climate change has been recognised relatively recently. Forest land and harvested wood products (HWPs) form the most important sink for greenhouse gases (GHGs) in the EU27. In 2018, the GHG net removal from forest land and the HWP sectors was −389 million tonnes of CO_2 equivalent, which corresponded to roughly 10% of the net emissions from all the sectors, excluding the land use, land use change and forestry (LULUCF) sector. The role of forests in achieving the goals of the Paris Agreement is also recognised in the LULUCF regulation ([EU] 2018/841), which aims to maintain and strengthen the forest sinks in the long term in order to reach the goal of balancing GHG emissions and removals in the second half of this century.

Until the financial crisis of 2007–2008, the GHG net removal (i.e. the carbon sink) of forest land fluctuated yearly in the EU27, but overall, the level of the forest sink was quite stable. After a temporary strengthening, the forest sink started to decline after 2013, however (Fig. Box 1.4). This decline is attributable to the age structure of the European forests, with the forests ageing and harvesting volumes increasing. The total volume of roundwood removals have been growing relatively steadily since 2009, while simultaneously, the share of wood fuel from total removals has increased slightly (Fig. Box 1.5). From the early 2000s up to the financial crisis, the share of wood fuel was one fifth of the total removal, whereas in the 2010s, the share was roughly a quarter. The increased production of wood fuel, as well as other short-lived wood-based products, such as pulp for paper and paperboard, is reflected in the GHG sink of HWPs that has not increased in parallel with the total roundwood removal volumes. In fact, the manufacture of long-lived wood products (i.e. sawnwood and panels) has only recently reached the pre-2009 level.

In forest-rich countries where the forest bioeconomy has more than a marginal role in the national economy, such as in Finland and Sweden, the national bioeconomy and forest strategies are aimed at increasing the use of woody biomass to reach the maximum sustainable volumes, alongside the nature conservation and biodiversity targets of the forests. At the European level, where the forest bioeconomy plays a minor role compared to agriculture and food manufacturing, the aims of the forest-rich countries are perhaps not fully understood. However, the recent forest damage due to storms and bark-beetle outbreaks in Central Europe, as well as the forest fires in the Mediterranean countries, Australia, Russia, California and Canada, have increased the general level of knowledge and understanding of the positive effects of active, sustainable forest management combined with a competitive, vibrant wood-processing industry in the context of climate change mitigation and adaptation.

References

Biber P, Felton A, Nieuwenhuis M, Lindbladh M, Black K, Bahýl J et al (2020) Forest biodiversity, carbon sequestration, and wood production: modeling synergies and trade-offs for ten Forest landscapes across Europe. Front Ecol Evol 8:e547696. https://doi.org/10.3389/fevo.2020.547696

D'Amato D, Droste N, Allen B, Kettunen M, Lähtinen K, Korhonen J et al (2017) Green, circular, bio economy: a comparative analysis of sustainability avenues. J Clean Prod 168:716–734. https://doi.org/10.1016/j.jclepro.2017.09.053

EEA (2021) Air emissions inventories. https://appsso.eurostat.ec.europa.eu/nui/show.do?dataset=env_air_gge&lang=en. Accessed 29 Apr 2021

European Commission (2012) Innovating for sustainable growth. A bioeconomy for Europe. https://op.europa.eu/en/publication-detail/-/publication/1f0d8515-8dc0-4435-ba53-9570e47dbd51. Accessed 9 Feb 2020

European Commission (2018) A sustainable bioeconomy for Europe: Strengthening the connection between economy, society and the environment. Updated Bioeconomy Strategy. https://ec.europa.eu/research/bioeconomy/pdf/ec_bioeconomy_strategy_2018.pdf#view=fit&pagemode=n one. Accessed 11 Feb 2020

European Environment Agency (2018) The circular economy and the bioeconomy. Partners in sustainability. EEA Report No. 8. ISSN 1977–844. https://www.eea.europa.eu/publications/circular-economy-and-bioeconomy. Accessed 10 Jan 2021

Eurostat (2021) GDP and its main components (output, expenditure and income). https://appsso.eurostat.ec.europa.eu/nui/show.do?dataset=nama_10_gdp&lang=en. Accessed 3 May 2021

FAO (2021) FAOSTAT. Forest product statistics. Forestry production and trade. http://www.fao.org/faostat/en/#data/FO. Accessed 30 Apr 2021

FAO (UN Food and Agricultural Organization) (2020) Global forest resource assessment. https://fra-data.fao.org/. Accessed 5 Jan 2021

FAO, ITTO, UN (2020) Forest product conversion factors. Rome. https://doi.org/10.4060/ca7952en. Accessed 30 Apr 2021

Helm D (2015) Natural capital: valuing the planet. Yale University Press, Yale

Hetemäki L (2019) The role of science in Forest policy – experiences by EFI. For Policy Econ 105:10–16. https://doi.org/10.1016/j.forpol.2019.05.014

Hetemäki L, Hanewinkel M, Muys B, Ollikainen M, Palahí M, Trasobares A (2017) Leading the way to a European circular bioeconomy strategy. From science to policy 5. European Forest Institute https://doi.org/10.36333/fs05

Hulme M (2009) Why we disagree about climate change: understanding controversy, inaction and opportunity. Cambridge University Press, Cambridge. https://doi.org/10.1017/CBO9780511841200

IPCC (2019) Climate change and land: an IPCC special report on climate change, 184 desertification, land degradation, sustainable land management, food security, and 185 greenhouse gas fluxes in terrestrial ecosystems. https://www.ipcc.ch/srccl/

JRC DataM (2021) Data portal of agro-economic research – DataM. https://datam.jrc.ec.europa.eu/datam/public/pages/index.xhtml. Accessed 30 Apr 2021

Kauppi P, Hanewinkel M, Lundmark L, Nabuurs GJ, Peltola H, Trasobares A, Hetemäki L (2018) Climate smart forestry in Europe. European Forest Institute. https://efi.int/publications-bank/climate-smart-forestry-europe

Kleinschmit D, Arts B, Giurca A, Mustalahti I, Sergent A, Pulzl H (2017) Environmental concerns in political bioeconomy discourses. Int For Rev Suppl 1. https://doi.org/10.1505/146554817822407420

Krumm F, Schuck A, Rigling A (eds) (2020) How to balance forestry and biodiversity conservation. A view across Europe 10.16904/envidat.196

Kuosmanen T, Kuosmanen N, El-Meligi A, Ronzon T, Gurria P, Iost S, M'Barek R (2020) How big is the bioeconomy? Reflections from an economic perspective. EUR 30167 EN, Publications

Office of the European Union, Luxembourg. https://publications.jrc.ec.europa.eu/repository/bitstream/JRC120324/how-big-is-the-economy.pdf. Accessed 20 June 2020

Nabuurs GJ, Delacote P, Ellison D, Hanewinkel M, Lindner M, Nesbit M, Ollikainen M, Savaresi A (2015) A new role for forests and the forest sector in the EU post-2020 climate targets. From science to policy 2. European Forest Institute. https://doi.org/10.36333/fs02

Nabuurs G-J, Delacote P, Ellison D, Hanewinkel M, Hetemäki L, Lindner M (2017) By 2050 the mitigation effects of EU forests could nearly double through climate smart forestry. Forests 8:484. https://doi.org/10.3390/f8120484

Nikolakis W, Innes J (eds) (2020) The wicked problem of Forest policy. Cambridge University Press, Cambridge. ISBN 9781108471404. https://doi.org/10.1017/9781108684439

Palahí M, Pantsar M, Costanza R, Kubiszewski I, Potočnik J, Stuchtey M, ... Bas L (2020) Investing in nature as the true engine of our economy: a 10-point action plan for a circular bioeconomy of wellbeing. Knowledge to action 02. European Forest Institute https://doi.org/10.36333/k2a02

Regulation (EU) (2018/841) Of the European Parliament and of the Council of 30 May 2018 on the inclusion of greenhouse gas emissions and removals from land use, land use change and forestry in the 2030 climate and energy framework, and amending Regulation (EU) No 525/2013 and Decision No 529/2013/EU (Text with EEA relevance). https://eur-lex.europa.eu/legal-content/EN/TXT/PDF/?uri=CELEX:32018R0841&from=EN. Accessed 11 Feb 2020

Robert N, Jonsson R, Chudy R, Camia A (2020) The EU bioeconomy: supporting an employment shift downstream in the wood-based value chains? Sustainability 2020(12):758. https://doi.org/10.3390/su12030758. Accessed 2 May 2021

Ronzon T, Piotrowski S, M'Barek R, Carus M (2017) A systematic approach to understanding and quantifying the EU's bioeconomy. Bio-Based Appl Econ 6(1):1–17. https://doi.org/10.13128/BAE-20567

Yousefpour R, Augustynczik ALD, Reyer C, Lasch P, Suckow F, Hanewinkel M (2018) Realizing mitigation efficiency of European commercial forests by climate smart forestry. Sci Rep 8:345. https://doi.org/10.1038/s41598-017-18778-w

Chapter 2
Planetary Boundaries and the Role of the Forest-Based Sector

Lauri Hetemäki and Jyri Seppälä

Abstract 'Planetary boundaries' is a concept that has been introduced by Earth system scientists to refer particularly to anthropogenic pressures on the Earth system that have reached a scale where abrupt global environmental change can no longer be excluded. In the planetary boundaries discussion, climate change plays a central role due to its overarching impacts on all the other planetary boundaries. For example, climate change critically impacts biodiversity and land-use changes. Consequently, climate change shapes policies, strategies and actions at the global, continental, national, regional and individual levels. The main policy through which the EU is seeking to address climate change and direct the region to live within the planetary boundaries is the European Green Deal (EGD), launched in 2019. The EGD clearly acknowledges the role forests can play in sinking carbon and suggests measures to enhance forest restoration and conservation. However, it falls short of recognising the role that the forest-based bioeconomy can also play in achieving the EGD objectives. History shows that European forests can simultaneously increase the carbon sink, biodiversity and wood production.

Keywords Planetray boundaries · Environment · Climate change · Green deal · Forest bioeconomy

2.1 The Period of Wakening to Planetary Boundaries

Humans have a tendency to see the times they are living in as periods of exceptional change, something that is historically very different from the past. Globalisation and the spread of the Internet at the turn of the century, and the financial crises of

L. Hetemäki (✉)
European Forest Institute, Joensuu, Finland

Faculty of Agriculture and Forestry, University of Helsinki, Helsinki, Finland
e-mail: lauri.a.hetemaki@helsinki.fi

J. Seppälä
Finnish Environment Institute, Helsinki, Finland

© The Author(s) 2022
L. Hetemäki et al. (eds.), *Forest Bioeconomy and Climate Change*, Managing
Forest Ecosystems 42, https://doi.org/10.1007/978-3-030-99206-4_2

2008–2010 are two such recent examples. In hindsight, these events changed many things significantly, and indeed could perhaps be seen as creating exceptional times. Currently, we seem to be facing yet another major periodic structural change, one that seems to be even more significant than the two we have already experienced in this century. Here, we identify this as the 'period of wakening to planetary boundaries'. This period has its roots in scientists' warnings, and is manifesting itself in increasing societal awareness of environmental concerns, and new international and national policy agendas directed at these. Time will tell how significant this period turns out to be, but currently the expectation is that it will lead to systemic changes in society, rather than only some fine-tuning. Here, we explain in more detail what we mean by this, and how it relates to the theme of this book.

'Planetary boundaries' is a concept that has been introduced by Earth system scientists (Rockström et al. 2009; Steffen et al. 2015; Otto et al. 2020). It refers to anthropogenic pressures on the Earth system that have reached a scale where abrupt global environmental change can no longer be excluded (Rockström et al. 2009). Accordingly, Rockström et al. (2009) proposed a new approach to global sustainability that defines planetary boundaries within which humanity can expect to operate safely. They identified nine planetary boundaries, including climate change and biodiversity, and argue that transgressing one or more planetary boundaries may be even catastrophic due to the risk of crossing thresholds that will trigger non-linear, abrupt environmental change within continental- to planetary-scale systems (Rockström et al. 2009). Moreover, according to Otto et al. (2020), technological progress and policy implementations are required to deliver emissions reductions at rates sufficiently fast to avoid crossing dangerous tipping points in the Earth's climate. Scientists and experts are also making suggestions for policy actions to avoid the tipping points. Palahí et al. (2020b) developed a 10-point action plan on how to respond to these challenges. The forest-based sector is understood to have an important role in helping to contribute to the solutions (Hetemäki et al. 2017; Palahí et al. 2020a, b).

This type of rhetoric concerning tipping-points and warnings is reminiscent of the 'limits-to-growth' debate of the 1970s. However, the limits-to-growth discussion emphasised the quantity of growth and the limits to the quantity of natural resources, whereas the planetary-boundaries discussion places emphasis on the quality of growth and the environmentally sustainable use of natural resources, as well as the need for circular economies and the mitigation of climate change, which were not major issues in the 1970s.

Nevertheless, it is evident that, globally, there has been a new type of awakening to environmental sustainability. People have reacted with heightened readiness, voiced their worries and taken action on climate change and biodiversity issues. In particular, the younger generations have become very active on these issues, and have managed to capture media attention and spread greater societal awareness, including to, it seems, politicians. This is evidenced by, for example, Greta Thunberg's school strike for climate action movement and the attention it created globally. The 'biodiversity crisis' that has loomed for a long time in the shadows of climate change discussions has recently been brought to a new level of societal awareness with the COVID-19 pandemic in 2020. Scientists have pointed out how these types of zoonotic diseases are linked to biodiversity, and why biodiversity loss is likely to make zoonotic diseases more frequent (Intergovernmental Science-Policy Platform on Biodiversity and Ecosystem Services 2020).

Politicians are being awakened to a new degree of seriousness and urgency on climate and biodiversity issues. Of course, these are not new to the political agenda; for example, the Kyoto Protocol international climate treaty was adopted over two decades ago, in 1997, and similarly, the Rio Conference in 1992 established the Convention on Biological Diversity. However, the last 30 years have witnessed insufficient, or even no, action to seriously change economic and societal structures to be in line with the goals of previous agreements. This has itself worsened sustainability development and made the rationale for the agreements even more urgent and important than three decades ago. Also, the scientific evidence pointing to the serious risks of transgressing the planetary boundaries has become stronger and broader. The bulk of the voting population, at least in the EU, is also starting to be increasingly concerned about the negative impacts of climate change in their everyday lives and to worry about the future. These changes have finally led also politicians to understand the importance of the issue and having a sense of urgency to act. Moreover, the majority of politicians are no longer talking about the need to fine-tune our economies and societies gradually to tackle these issues, but are increasingly calling for systemic and urgent changes (e.g. the European Green Deal [EGD]).

The 'period of wakening to planetary boundaries' can be seen in the Paris Climate Agreement and Sustainable Development Goals, which the world's states agreed to in 2018. Since then, these agreements have been beacons for national and regional strategies and policies, more or less everywhere around the globe. However, as one would expect, there has been variation in how strongly these agreements have been realised in new policy measures. For example, the USA pulled out of the Paris Agreement during ex-President Trump's period in office (although the USA rejoined at the start of President Biden's term), whereas the EU aims to implement the main goals via its EGD programme, launched in December 2019. In general, the trend of viewing the environment as a major priority in political agendas appears to be becoming stronger in an increasing number of world regions, and day-by-day. One of the latest examples is President Xi Jinping's announcement at the UN General Assembly in 2020 that China's emissions will peak before 2030 and they will strive to reach carbon neutrality before 2060. Whether these goals will be achieved is still to be seen, but nevertheless, the political goal is set.

2.2 Climate Change as the Deciding Phenomenon

In the planetary-boundaries discussion, climate change plays a central role due to its overarching impacts on all the other (eight) planetary boundaries. For example, climate change critically impacts biodiversity and land-use changes. Due to the drastic consequences of climate change, there is no question that it will be the deciding phenomenon of our times. It is shaping policies, strategies and actions at the global, continental, national, regional and individual levels.

Moreover, the science fundamentals clearly indicate that greenhouse gases (GHGs) caused by humans have been the dominant influence on the climate system at least since the twentieth century. The Intergovernmental Panel on Climate Change (IPCC 2018) concluded that human-induced warming has exceeded 1 °C above pre-industrial levels, and is continuing to increase at a rate of 0.2 °C per decade. Human-induced warming will exceed 1.5 °C around 2040, if this rate of increase continues. To avoid this development, 190 countries signed the Paris Agreement (United Nations [UN] 2020). This sets out a global framework to avoid dangerous climate change by limiting global warming to well below 2 °C and pursuing efforts to limit it to 1.5 °C. Although nearly all the countries have ratified the Paris Agreement, the curtailing of global greenhouse emissions has not proceeded so far according to pathways that align with the Paris goal (International Energy Agency 2020a, b).

Despite scientific consensus about the role of anthropogenic GHG emissions on climate change, there is uncertainty in the climate response to emissions of GHGs. It is only possible to provide probabilities of the impacts of different emissions pathways on climate warming up to a certain point (e.g. below 2 °C). In emissions pathway scenarios, the reduction of long-lived CO_2 emissions is the priority because cumulative CO_2 emissions are the main determinant of future warming. Nitrous oxide and fluorinated gases are also long-lived GHG emissions, but their absolute global warming impacts are much lower compared to CO_2 due to their emissions being magnitudes smaller. Short-lived GHGs, such as methane, affect the climate in different ways compared to long-lived GHG emissions. The constant rates of their emissions do not lead to increasing warming, whereas CO_2 emissions do. Despite their differences, the various GHG emissions are typically aggregated as 'carbon dioxide equivalence', describing their 100-year time-horizon warming impact relative to CO_2.

Human activities are estimated to have caused approximately 1.1 °C of global warming above pre-industrial levels, and the six warmest years on record have taken place since 2014 (NASA 2020). The ocean has absorbed much of the increased heat, with the top 100 m warming by more than 0.33 °C since 1969 (von Schuckmann et al. 2020). Warmer waters, and melting water from ice-sheet mass loss in Greenland and Antarctica, have increased ocean sea levels (Velicogna et al. 2020). The global sea level rose about 20 cm in the last century. However, the rate in the last two decades is nearly double that of the last century, and is accelerating slightly every year (Nerem et al. 2018). It is estimated that the sea level will rise by 30–240 cm by 2100, depending on developments in lowering global GHG emissions in the future (IPCC 2019). Global climate change has already had many observable effects on the environment: glaciers have shrunk, the ice cover over the Arctic Ocean has decreased, the frost-free season (and growing season) in the north has lengthened, changes in precipitation patterns, and more extreme meteorological events, such as hurricanes, storms, droughts, floods and heat waves, have been detected in different parts of the world (NASA 2020).

According to the IPCC (2018), the extent of the effects of climate change on individual regions will vary over time and according the ability of different societal and environmental systems to mitigate or adapt to the change. Almost all coastal cities of any size, and all small, developing island states, are becoming increasingly vulnerable to rising sea levels (UN 2019). In particular, developing countries will suffer from more frequent and extreme weather events caused by climate change, and they will have more problems with food security and water scarcity in the future. In addition to the harmful effects on human systems, including human health, climate change will cause negative impacts on natural systems—air, biological diversity, freshwater, oceans and land—which will alter the complex interactions between the human and natural systems (UN 2019). For this reason, the key message of the IPCC special report on the impacts of global warming by 1.5 °C was that the harmful effects of climate change will increase so rapidly between the targets of 1.5 and 2.5 °C that it is imperative to limit global warming to 1.5 °C.

In order to stop global temperatures increasing, global emissions of CO_2 gases must be cut to near net-zero by around the middle of this century in most 1.5 °C scenarios, and around 2075 for 'well below' 2 °C scenarios. In addition, net-zero GHG emissions in these scenarios must be typically reached around 15 years later than reaching net-zero CO_2 emissions. Both net-zero situations would require the large-scale net removal of CO_2 from the atmosphere because all anthropogenic CO_2 emissions, and especially non-CO_2 emissions, may not be stopped in the future. The removal of CO_2 can be implemented, for example, with the help of afforestation and carbon- removal technologies, such as bioenergy with carbon capture and storage.

Chapter 3 contains a more-detailed discussion of the recent IPCC projections for global climate change. There is also an analysis on what this could mean, especially to boreal forests. However, next we turn to examining how the EU policy framework—the EGD—addresses the forest-based sector in climate mitigation.

2.3 The European Green Deal and the Forest-Based Sector

The EU aims to implement the main goals of the Paris Agreement via the EGD: "…a new growth strategy that aims to transform the EU into a fair and prosperous society, with a modern, resource-efficient and competitive economy where there are no net emissions of greenhouse gases in 2050 and where economic growth is decoupled from resource use. It also aims to protect, conserve and enhance the EU's natural capital, and protect the health and well-being of citizens from environment-related risks and impacts (European Commission [EC] 2019)". In essence, climate change mitigation and biodiversity will be at the centre of EU policies in the years to come.

The EGD policy document is, in many ways, a landmark, representing a new way of thinking in the EC. It is aimed at being a cross-sectoral policy outline that will have an effect on all legislative processes of the EC in 2020–2024. The political importance of the EGD is also evident in the requirement that "All EU actions and policies will have to contribute to the European Green Deal objectives" (EC 2019, p. 3). This includes many EU forest-sector-related areas, such as climate policy, biodiversity policy, energy policy, forest strategy, industrial policy, etc. The implementation of the strategies and polices proposed in the EGD will have significant implications for the EU forest sector in the coming decade. The EGD introduces a new political narrative and direction by setting a clear focus on climate, sustainability and biodiversity conservation across all policy areas. The EGD acknowledges the need for a systemic transformation, not only piecemeal policy changes, to achieve the goals set by the Paris Climate Agreement, Sustainable Development Goals and Convention on Biological Diversity.

The main goal of the EGD is for the EU to become the world's first climate-neutral continent. To reach this goal, European GHG emissions and sinks should be equal in 2050. In addition to the fossil- and process-based GHG emissions included in the EU Emissions Trading System (EU ETS) and the EU's Effort Sharing Decision (non-ETS), land-based emissions and sinks that occur in the land-use, land-use change and forestry (LULUCF) sector are being considered as new elements for EU climate policy. The so-called 'no-debit rule'—a principle applied in EU law for the first time for 2021–2030—requires that GHG emissions from the LULUCF sector are compensated for by an equivalent absorption of CO_2 made possible by additional action in this sector (EU 2018). The absorption can be effected through carbon sinks in agricultural soils and, especially, forest-related sinks. Thus, the actions of forest owners and farmers to secure carbon stored in forests and soils will contribute to achieving the EU's climate-neutral target by 2050.

The EGD clearly acknowledges many of the potential problems relating to forests. Most of its statements regarding forests express problems such deforestation and threats to forests and biodiversity, and argue for forest and biodiversity restoration and protection. With respect to climate action, forests are mainly viewed as carbon sinks. There are hardly any statements on the multiple benefits forests provide to society, or the benefits that forest-based bioindustry could contribute to a more sustainable and climate-neutral society and to the Sustainable Development Goals. Indeed, Palahí et al. (2020a, b) argued that the bioeconomy is *the missing link* in the EGD, stating that "The bioeconomy, a circular economy based on renewable biological resources and sustainable biobased solutions, could certainly contribute to the Green Deal delivery and would deserve more attention. The bioeconomy can be a catalyst for systemic change to tackle holistically the social, economic and environmental aspects currently not yet enough coherently addressed". A sustainably managed forest bioeconomy— sustainability not just assumed, but imposed and monitored—could deliver the following EGD objectives.

First, moving towards a carbon-neutral EU requires not only moving towards fossil-free energy, but also to fossil-free materials. This means replacing carbon-intense products, such as plastics, concrete, steel and synthetic textiles. This is not only for climate change mitigation, but also because of other positive environmental impacts. The transformation called for in the EGD is simply not possible without using a new range of renewable biobased materials that can replace and environmentally outperform carbon-intense materials. This shift also provides an opportunity for modernisation and making industries more circular. Forest resources, if managed sustainably, are circular by nature and often easy to remanufacture. The EGD identifies several sectors, such as chemicals, textiles, plastics and construction, which will need new conceptual business models and innovations to become circular and low-carbon industries. The emerging bioeconomy could be a catalyst for this. Wood—the most versatile biological material on Earth—can be transformed into nanocellulose, which is five times stronger and lighter than steel. The first car made of nanocellulose was unveiled in 2019 in Japan. A new generation of sustainable and circular wood-based textiles, with much lower carbon footprints than fossil fibres like polyester, is now possible too (Hurmekoski et al. 2018). Engineered-wood products, such as cross-laminated timber elements and modules, are the most effective way to reduce the carbon footprint in cities and the construction sector, both of which are currently dominated by carbon- and resource-intense materials—concrete and steel.

Second, the bioeconomy offers an opportunity to address the past failure of the economy to *value nature and biodiversity*. This is because a sustainable bioeconomy needs to place nature and life at the centre of the economy. Biological diversity determines the capacity of biological resources to adapt and evolve in a changing environment. Biodiversity is therefore a prerequisite for a long-term, sustainable and resilient bioeconomy. On the other hand, a sustainable bioeconomy is necessary in the long term to protect biodiversity, as new biobased solutions to replace fossil products are crucial in mitigating climate change, biodiversity's main threat. Moreover, forest management, such as promoting more resilient mixed forests or addressing natural disturbances, can simultaneously benefit biodiversity and the bioeconomy (Biber et al. 2020; Díaz-Yáñez et al. 2020; Krumm et al. 2020). Third, it is important to acknowledge that it is unlikely that actions to protect or enhance biodiversity can be funded by public money only. Forest owners and forest industries, generating enough income from a profitable bioeconomy, would be in a better position to reinvest in biodiversity and natural capital, in line with the aims of the EGD of *preserving and restoring ecosystems and biodiversity*.

Finally, the bioeconomy offers unique opportunities for inclusive prosperity and fair social transition. This is paradoxically related to one of the potential disadvantages of the bioeconomy compared to the fossil-based economy—a more complex ownership, mobilisation and processing of biological resources. Biological resources like forest resources are usually owned by many more people and entities, they are located in diverse rural areas, their costs are often higher, and transporting and processing biomass tends to be more costly and complex compared to fossil resources, such as coal and oil. However, these limitations are, at the same time, a great advantage, as they offer the possibility of a more-inclusive distribution of income, jobs, infrastructure and prosperity in many regions of the EU, especially in rural areas, in line with the EGD's inclusive growth ambitions. For instance, forests cover more than 40% of the EU land surface, and the forest-based sector provides 3.5 million jobs. This is more than the three top energy-intensive industries (steel, chemicals and cement), which the EGD called "indispensable to Europe's economy", while forgetting to even mention the forest industry. In addition, the EU forest-based sector also includes 400,000 small- and medium-scale enterprises and 16 million forest owners. Thus, the forest-based sector offers an extensive and unique socioecological 'fabric' in which to progress the EGD ambitions.

In summary, the bioeconomy, when managed in a sustainable way, has major potential for helping to deliver the ambition set by the EGD. Palahí et al. (2020a, b) also stated that "it is still an important missing part of the complicated puzzle to overcome the past dichotomy between economy and ecology that very much defined the twentieth century. The bioeconomy provides us with the opportunity to build a new and synergistic relationship between technology and nature, between ecology and economy that can define the twenty-first century: the century where we would finally start respecting the laws of physics and integrate biology".

2.4   Diverse Role of Forests and Forest-Based Products

As indicated above, the EDG views EU forests as mainly contributing to climate mitigation via forest sinks. Accordingly, the policy suggestions of the EDG emphasise reforestation, the restoration of degraded forests and forest conservation. The role of the forest bioeconomy in this effort is missing from the document. As argued above, this is a clear shortcoming, and hopefully it will be addressed later, in the design and implementation phase of the policies.

It is fitting to reflect on the importance of the global and EU forest-based sectors in climate mitigation based on some key statistics. Forests and wood are not equally distributed across world regions, and forests are also managed and used in different ways. These differences partly explain why forests have played diverse roles for nations, and also why cultural meaning and citizen perceptions of forests may differ across countries.

More than half (54%) of the world's forests are located in only five countries—the Russian Federation, Brazil, Canada, the USA and China (Food and Agriculture Organization [FAO] 2020). The forest area in the EU is relatively small in global terms, but its role as a forest bioeconomy products producer is a major one (Fig. 2.1). The EU forest area (ha) in 2020 accounted for only 3.9% of the world total, but its export value for forest products was 41% of the world total, amounting to USD95 billion (FAOSTAT). Although most of this export value figure is related to the EU's internal trade, the exports to regions outside the EU are also major. In 2017, the EU27 forest products export value to regions outside the EU27 was USD36.5 billion, or 37% of the total EU27 forest products export value. This was more than the combined forest products export value of Brazil, China and the Russian Federation (USD35.3 billion), whose share of the world forests is 38% (i.e. 10 times more than the EU27).

Figure 2.1 indicates that the potential impact of the EU's forest sink at the global level is low, and will be so in the future due to its small proportion of forest area. However, it is evident that the EU has to do its share in contributing to climate mitigation by enhancing forest sinks, and also providing an example of how this can be done via forest management. Its recent record on this has been quite good. According to the FAO, the forest area in Europe (excluding Russia) has increased by 17 million ha (9%) over the last three decades—an amount equal to the entire forest area of

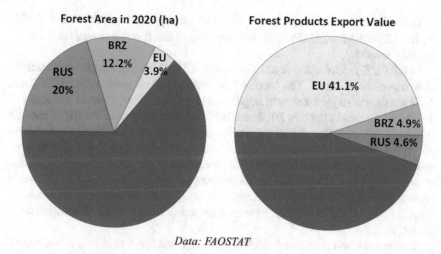

Data: FAOSTAT

Fig. 2.1 Shares of forest area and forest-product export values of the global total in 2020. (Data: FAOSTAT)

France. The volume of wood stock has increased even more, by 46%. Moreover, the carbon stock of forests in Europe (excluding Russia) has also been growing steadily over the last three decades—in 2020 it was 24% higher than in 1990 (FAO 2020).

On the other hand, the highly significant position the EU has in global forest-products exports indicates that it can play a major role in helping to advance the replacement of fossil-based energy, raw materials and products, and generally enhance sustainable production and consumption. To strengthen this role, EU27 forest bioeconomy product innovations, and increasing resource efficiency and circularity, will be key priorities. In doing this, the EU27 can also have a significant global impact in the movement to more sustainable production.

Interestingly, the increasing forest area and carbon sink in Europe has evolved simultaneously with a significant increase in wood production. Europe's (excluding Russia) timber production added up to 13.5 billion m³ between 1992 and 2019, increasing by 67% during this period. Consequently, Europe has concurrently increased wood production, forest area, the volume of wood stock and the extent of the carbon sink. Unfortunately, this trend has not been seen globally, with deforestation taking place over the last few decades, especially in Africa and South America.

According to the FAO, the area of protected forests in Europe has also more than doubled during the last three decades. However, they cover less than 8% of the area used for wood production. Clearly, as also suggested by the EU's Biodiversity Strategy from 2020, there is a need to further increase forest biodiversity. One important implication of the above data is, however, that it has been possible to enhance many of the forest ecosystem services at the same time. The EU should build on the lessons learnt from this history to implement its EGD.

But will there be enough wood for bioeconomy development in the EU27? Figure 2.2 shows the EU27 roundwood production and forest growing stock for the last two decades.[1]

In the EU27, roundwood production increased by 21% and the growing stock by 25% from 2000 to 2019. The 'outliers' in the roundwood production in 2005 and 2007 are due to large windstorm impacts, whereas the 2009 slump is the result of the global financial crisis. In 2018 and 2019, salvage logging in the EU27 appears to have been higher than the long-term average. Despite the increase in roundwood production, the EU27 produces slightly less wood today than it did two decades ago, in terms of the volume of trees in the forests. In 2019, 1.8% of the total volume of wood in the forests was used for roundwood production. Also, EU27 roundwood imports are today at the same level as they were at the beginning of this century. Thus, so far, wood resources have not impeded bioeconomic development in the EU27.

The future wood potential from EU27 forests, and the demand for it, are uncertain. The future wood supply potential will depend on factors such as the age structure of the forests, forest management measures, climate change impacts (positive and negative), afforestation and deforestation, the scale of conservation, etc. On the other hand, the demand for roundwood in the future will depend, for example, on the emergence of new bioproducts, the competitiveness of EU27 forest industries, better wood resource efficiency (e.g. using forest residues and production side

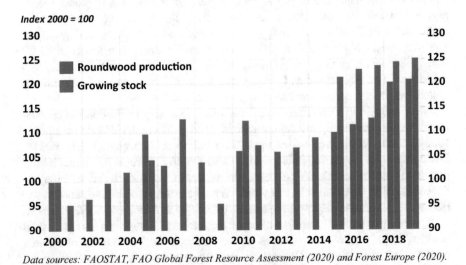

Data sources: FAOSTAT, FAO Global Forest Resource Assessment (2020) and Forest Europe (2020).

Fig. 2.2 EU27 roundwood production and forest growing stock, 2000–2019. (Data sources: FAOSTAT, FAO Global Forest Resource Assessment (2020) and Forest Europe (2020))

---

[1] The growing stock is the volume of all living trees in a given area of forest and it is a close approximation of the total aboveground biomass in forests. It forms the basis, for instance, for quantifying the carbon sequestration in forests. It usually involves measuring all the trees with a total height greater than 1.3 m.

streams), the declining demand for some traditional forest products (e.g. paper), and the level of roundwood imports and exports. What is worrying is that there is a lack of systematic global and European outlook studies that have comprehensively analysed the wood supply and demand in a way that the ongoing structural changes are taken into account (Hetemäki and Hurmekoski 2016; Jonsson et al. 2017; Hurmekoski et al. 2018; Hetemäki et al. 2020). Clearly, this is a serious shortcoming for the future planning and implementation of the EGD as well.

To conclude, motivated by the above facts, this book's focus is on the whole forest-based sector—not just forest sinks—when discussing the role of forests in climate change mitigation.

References

Biber P, Felton A, Nieuwenhuis M, Lindbladh M, Black K, Bahýl J et al (2020) Forest biodiversity, carbon sequestration, and wood production: modeling synergies and trade-offs for ten Forest landscapes across Europe. Front Ecol Evol 8:e547696. https://doi.org/10.3389/fevo.2020.547696

Committee on Climate Change (2019) Net zero: the UK's contribution to stopping global warming. Committee on Climate Change, London

Communiqué Global Bioeconomy Summit (2015) Making bioeconomy work for sustainable development. http://gbs2015.com/fileadmin/gbs2015/Downloads/Communique_final.pdf. Accessed 10 Jan 2021

Díaz-Yáñez O, Pukkala T, Packalen P, Lexer M, Peltola H (2020) Multi-objective forestry increases the production of ecosystem services. Forestry Int J For Res 2020:1–9. https://doi.org/10.1093/forestry/cpaa041

EU (2018) Regulation (EU) 2018/841 of the European Parliament and of the Council of 30 May 2018 on the inclusion of greenhouse gas emissions and removals from land use, land-use change and forestry in the 2030 climate and energy framework, and amending Regulation (EU) No. 525/2013 and Decision No. 529/2013/EU

European Commission (EC) (2019) The European Green Deal. Brussels, 11.12.2019, COM(2019) 640 final. https://eur-lex.europa.eu/legal-content/EN/TXT/?uri=COM%3A2019%3A640%3AFIN

FAOSTAT (2020). http://www.fao.org/faostat/en/#data/FO. Accessed XX Jan 2021

Forest Europe (2020) State of Europe's Forests 2020. https://foresteurope.org/state-europes-forests-2020

Griggs D, Stafford-Smith M, Gaffney O, Rockström J, Öhman MC, Shyamsundar P et al (2013) Sustainable development goals for people and planet. Nature 495:305–307

Hetemäki L, Hurmekoski E (2016) Forest products markets under change: review and research Implications. Curr For Rep 2:177–188. https://doi.org/10.1007/s40725-016-0042-z

Hetemäki L, Hanewinkel M, Muys B, Ollikainen M, Palahí M, Trasobares A (2017) Leading the way to a European circular bioeconomy strategy. From Science to Policy 5. European Forest Institute. https://doi.org/10.36333/fs05

Hetemäki L, Palahi M, Nasi R (2020) Seeing the wood in the forests. Knowledge to Action no. 1. European Forest Institute. https://doi.org/10.36333/k2a01

Hulme M (2009) Why we disagree about climate change: understanding controversy, inaction and opportunity. Cambridge University Press, Cambridge. https://doi.org/10.1017/CBO9780511841200

Hurmekoski E, Jonsson R, Korhonen J, Jänis J, Mäkinen M, Leskinen P, Hetemäki L (2018) Diversification of the forest industries: role of new wood-based products. Can J For Res 48(12):1417–1432. https://doi.org/10.1139/cjfr-2018-0116

IEA (2020a) Coal information: overview. Statistics report, July 2020. https://www.iea.org/reports/coal-information-overview. Accessed XX Jan 2021

IEA (2020b) Global CO_2 emissions in 2019. https://www.iea.org/articles/global-co2-emissions-in-2019. Accessed XX Jan 2021

Intergovernmental Platform on Biodiversity and Ecosystem Services (IPBES) (2020) IPBES workshop on biodiversity and pandemics: executive summary. https://ipbes.net/. Accessed XX Jan 2021

IPCC (2018) Special report on global warming of 1.5°C approved by governments. IPCC, Switzerland

IPCC (2019) The ocean and cryosphere in a changing climate – a special report of the intergovernmental panel on climate change. IPCC, Switzerland

Jonsson R, Hurmekoski E, Hetemäki L, Prestemon J (2017) What is the current state of forest product markets and how will they develop in the future? In: Winkel G (ed) Towards a sustainable European forest-based bioeconomy – assessment and the way forward. What science can tell us, no. 8. European Forest Institute. https://efi.int/sites/default/files/files/publication-bank/2018/efi_wsctu8_2017.pdf

Krumm F, Schuck A, Rigling A (Eds.) (2020) How to balance forestry and biodiversity conservation. A view across Europe. European Forest Institute and Swiss Federal Research Institute. https://doi.org/10.16904/envidat.196

Lindner M, Maroschek M, Netherer S, Kremer A, Barbati A, Garcia-Gonzalo J et al (2010) Climate change impacts, adaptive capacity, and vulnerability of European forest ecosystems. For Ecol Manag 259:698–709. https://doi.org/10.1016/j.foreco.2009.09.023

Matson PA, Parton WJ, Power AG, Swift MJ (1997) Agricultural intensification and ecosystem properties. Science 277(5325):504–509. https://doi.org/10.1126/science.277.5325.504

Michud A, Tanttu M, Asaadi S, Ma Y, Netti E, Kääriainen P,… Sixta H (2016) Ioncell-F: ionic liquid-based cellulosic textile fibers as an alternative to viscose and Lyocell. Text Res J 86:543–552

Millennium Ecosystem Assessment (2005) Ecosystems and human well-being: synthesis. Island Press, Washington, DC

Nabuurs G-J, Delacote P, Ellison D, Hanewinkel M, Lindner M, Nesbit M, Ollikainen M, Savaresi A (2015) A new role for forests and the forest sector in the EU post-2020 climate targets. From Science to Policy 2. European Forest Institute. 10.36333/fs02

Nabuurs G-J, Delacote P, Ellison D, Hanewinkel M, Hetemäki L, Lindner M, Ollikainen M (2017) Mitigation effects of EU forests could nearly double by 2050 through climate smart forestry. Forests 8(12):484. https://doi.org/10.3390/f8120484

NASA (2020) Climate change: how do we know? https://climate.nasa.gov/evidence/. Accessed 28 Nov 2020

Nerem RS, Beckley BD, Fasullo JT, Hamlington BD, Masters D, Mitchum GT (2018) Climate-change–driven accelerated sealevel rise detected in the altimeter era. PNAS 115(9):2022–2025. https://doi.org/10.1073/pnas.1717312115

Official Journal of the European Union (2011) Regulation (EU) No 305/2011 of the European Parliament and of the Council of 9 March 2011. http://eur-lex.europa.eu/legal-content/EN/TXT/PDF/?uri=CELEX:32011R0305&from=EN. Accessed 20 Jan 2021

Ollikainen M (2014) Forests in the bioeconomy – smart green growth for humankind. Scand. J For Res 29:360–366

Otto IM, Donges JF, Cremades R, Bhowmik A, Hewitt RJ, Lucht W et al (2020) Social tipping dynamics for stabilizing Earth's climate by 2050. PNAS 117(5):2354–2365. https://doi.org/10.1073/pnas.1900577117

Palahí M, Hetemäki L, Potocnik J (2020a) Bioeconomy: The missing link to connect the dots in the EU Green Deal. EURACTIVE. https://pr.euractiv.com/pr/bioeconomy-missing-link-connect-dots-eu-green-deal-202385 and European Forest Institute blog: https://blog.efi.int/bioeconomy-the-missing-link-to-connect-the-dots-in-the-eu-green-deal/ Accessed 20 Mar 2020

Palahí M, Pantsar M, Costanza R, Kubiszewski I, Potočnik J, Stuchtey M, … Bas L (2020b) Investing in Nature as the true engine of our economy: A 10-point Action Plan for a Circular Bioeconomy of Wellbeing. Knowledge to Action 02, European Forest Institute. https://doi.org/10.36333/k2a02

Priebe J, Mårald E, Nordin A (2020) Narrow pasts and futures: how frames of sustainability transformation limit societal change. J Environ Stud Sci 11(2). https://doi.org/10.1007/s13412-020-00636-3

Priefer C, Jörissen J, Frör O (2017) Pathways to shape bioeconomy. Resources 6:1–23

Rockström J, Steffen W, Noone K, Persson Å, Chapin FS III, Lambin E et al (2009) Planetary boundaries: exploring the safe operating space for humanity. Ecol Soc 14(2):32. http://www.ecologyandsociety.org/vol14/iss2/art32/

Rockström J, Gaffney O, Rogelj J, Meinshausen M, Nakicenovic N, Schellnhuber HJ (2017) A roadmap for rapid decarbonization. Science 355(6331). https://doi.org/10.1126/science.aah3443

Steffen W, Richardson K, Rockström J, Cornell SE, Fetzer I, Bennett EM et al (2015) Planetary boundaries: guiding human development on a changing planet. Science 347(6223). https://doi.org/10.1126/science.1259855

Sukhdev PW, Schröter-Schlaack H, Nesshöver C, Bishop C, Brink J (2010) The economics of ecosystems and biodiversity: mainstreaming the economics of nature: a synthesis of the approach, conclusions and recommendations of TEEB (No. 333.95 E19). UNEP, Geneva

United Nations (2019) Global environment outlook GEO-6 – healthy planet, healthy people. United Nations Environment Programme. Cambridge University Press, Cambridge

United Nations (2020) Paris agreement – status of ratification. https://unfccc.int/process/the-paris-agreement/status-of-ratification. Accessed 26 Apr 2020

Velicogna I, Mohajerani Y, Geruo A, Landerer F, Mouginot J, Noel B et al (2020) Continuity of ice sheet mass loss in Greenland and Antarctica from the GRACE and GRACE follow-on missions. Geophys Res Lett 47(8):e2020GL087291. https://doi.org/10.1029/2020GL087291

von Schuckmann K, Cheng L, Palmer D, Hansen J, Tassone C, Aich V et al (2020) Heat stored in the Earth system: where does the energy go? Earth Syst Sci Data 12(3):2013–2041. https://doi.org/10.5194/essd-12-2013-2020

William J, Ripple WJ, Wolf C, Galetti M, Newsome TM, Alamgir M, Crist E, Mahmoud MI, Laurance WF (2017) World scientists' warning to humanity: a second notice. Bioscience 67(12):1026–1028

Chapter 3
Climate Change, Impacts, Adaptation and Risk Management

Ari Venäläinen, Kimmo Ruosteenoja, Ilari Lehtonen, Mikko Laapas, Olli-Pekka Tikkanen, and Heli Peltola

Abstract Under the moderate future greenhouse gas emissions scenario (RCP4.5), climate model simulations project that the annual mean temperature will increase in Europe by up to 2–3 °C by the middle of this century, compared to the end of the nineteenth century. The temperature increase is projected to be larger in Northern Europe than in Central and Southern Europe. The annual precipitation is projected to decrease in Southern Europe and increase in Northern and Central Europe. The projected changes in temperature and precipitation are expected to be higher in the winter than in the summer months. In Northern Europe, forest growth is generally projected to increase due to warmer and longer growing seasons. In southern Europe in particular, warmer and dryer summers are projected to decrease forest growth. Climate change is expected also to expose forests and forestry to multiple abiotic and biotic risks throughout Europe. The greatest abiotic risks to forests are caused by windstorms, drought, forest fires and extreme snow loading on trees. The warmer climate will also increase biotic risks to forests, such as damage caused by European spruce bark beetle (*Ips typographus*) outbreaks in Norway spruce (*Picea abies*) forests and wood decay by *Heterobasidion* spp. root rot in Norway spruce and Scots pine (*Pinus sylvestris*) forests. Different adaptation and risk management actions may be needed, depending on geographical region and time span, in order to maintain forest resilience, which is also important for climate change mitigation.

Keywords Adaptive management · European forests · Forest resilience · Global greenhouse gas emission scenarios · Impacts of climate change · Natural disturbances

A. Venäläinen (✉) · K. Ruosteenoja · I. Lehtonen · M. Laapas
Finnish Meteorological Institute, Helsinki, Finland

O.-P. Tikkanen · H. Peltola
University of Eastern Finland, Joensuu, Finland

© The Author(s) 2022
L. Hetemäki et al. (eds.), *Forest Bioeconomy and Climate Change*, Managing
Forest Ecosystems 42, https://doi.org/10.1007/978-3-030-99206-4_3

3.1 Global Climate Change

3.1.1 Global Climate Change in the Past

During the four billion years of planet Earth's existence, global climate has fluctuated greatly. Basically, these variations have been controlled by the heat balance of the planet. Virtually all the energy that drives the climate system originates from the sun. Approximately 70% of the total incoming solar radiation is absorbed by the earth, whilst the remaining 30% is reflected back into space. The energy input from solar radiation is balanced by the emission of thermal infrared radiation into space. A major part of the thermal radiation emitted by the surface is absorbed and then re-emitted by the atmosphere before ending up in space. This phenomenon is known as the greenhouse effect. The effectiveness of the phenomenon depends on the concentrations of various gases in the atmosphere. The most important greenhouse gases are water vapour and carbon dioxide (CO_2). Methane, ozone, nitrous oxide and several other gases likewise have some importance. In the absence of the greenhouse effect, the average surface temperature on Earth would be about -18 °C, whereas the actual current global mean is $+14$ °C; that is, more than 30 °C higher than without the greenhouse effect.

Natural climate changes in the history of the earth have been caused by multiple factors. These include long-term variations in the solar radiance, changes in the composition of the atmosphere, continental drifts and volcanic eruptions. During the past few million years, the climate has mainly been relatively cool, and ice ages with milder interglacial periods have followed one another on time scales of 10,000–100,000 years. Such glacial–interglacial variations are primarily induced by changes in Earth's orbit and axis of rotation. In addition, such variations are amplified by synchronous shifts in atmospheric CO_2 concentrations.

Since the pre-industrial era (i.e. from the nineteenth century), the global mean temperature has increased by about 1 °C. Accordingly, over the last few centurieis and decades, global climate has changed very rapidly compared to the trends typically experienced over millions of years in the past. This is due to the large increase in human-induced emissions of greenhouse gases — especially CO_2 — into the atmosphere. The major source of anthropogenic CO_2 emissions is the combustion of fossil fuels, the use of which has increased tremendously in tandem with global energy consumption. Deforestation and other changes in land use have also contributed to such emissions, albeit to a lesser extent. During the 1980s, climate change became recognised as a serious challenge to humankind. In order to respond to this challenge, the United Nations (UN) endorsed the establishment of an Intergovernmental Panel on Climate Change (IPCC) in 1988 (Box 3.1).

Box 3.1 Intergovernmental Panel on Climate Change (IPCC)
The United Nations (UN) endorsed the establishment of IPCC during its General Assembly in 1988. The IPCC was set up under the UN Environment Programme (UNEP) and the World Meteorological Organization (WMO). The IPCC is an organisation of the governments that are members of the UN, with the number of members currently being 195. The objective of the IPCC is to provide state-of-the-art scientific information about climate change. Besides climate change as a phenomenon, the IPCC produces comprehensive reviews and recommendations about the social and economic impacts of such change, along with potential response strategies. Consequently, the IPCC plays a fundamental role in the international conventions on climate. Since 1988, the IPCC has published five comprehensive assessment reports concerning climate change. The Fifth Assessment Report (AR5), finalised in 2013–2014, provided the scientific input for the Paris Agreement. Further the Sixth Assessment cycle has provided three Special Reports, a Methodology Report and the Sixth Assessment Report. The first Special Report — '*Global warming of 1.5 °C*' — was requested by world governments under the Paris Agreement, and was published in October 2018. The '*Special Report on Climate Change and Land*' (SRCCL) was published in August 2019, and the '*Special Report on the Ocean and Cryosphere in a Changing Climate*' (SROCC) in September 2019.

The concentration of atmospheric CO_2 has increased from a pre-industrial level of 280 ppm (parts per million by volume) to about 410 ppm in 2019. Simultaneously, the concentration of methane has more than doubled. On the other hand, a concurrent increase in the amount of sulphates and other aerosol particles originating from anthropogenic emissions has partially compensated for the warming effect from the increasing greenhouse gas concentrations. Aerosol-induced cooling results from the increased reflectance of solar radiation back into space. Nevertheless, greenhouse gas concentrations will eventually overwhelm this phenomenon because they continue to accumulate in the atmosphere, whereas aerosol particles are continuously being washed out from the atmosphere. Current emissions of CO_2 will be influencing the atmospheric composition for several millennia.

3.1.2   Assessment of Future Global Climate Change

Projections of future climates are derived from simulations performed using global climate models (GCMs). These models simulate the behaviour of the climate system by means of the application of physical laws. Climate models include discrete components for the atmosphere, oceans, soil, vegetation and cryosphere, and also consider interactions among these subsystems. In assessing future climate, such models are forced using different atmospheric greenhouse gas concentration scenarios (Box 3.2). Climate models require large computational resources, and thus they need to be run on supercomputers. Even so, the available computational capacity does not allow the models to simulate all the processes of the climate system in full detail. Hence, simplifying approximations are necessary, and these simplifications are implemented in different ways in the various models. Consequently, simulated future climatic changes diverge among the models. To obtain the most realistic picture of anticipated future changes and their uncertainties, it is recommended to use a wide array of climate models rather than rely on only one or a few.

Figure 3.1 illustrates the global emissions and atmospheric concentrations of CO_2, as well as the modelled evolution of mean global warming, under three RCP scenarios (for further information about the RCP scenarios, see the Box 3.2). The changes in temperature are given relative to the temporal mean of the period 1971–2000. Prior to this period, the global mean temperature had already risen by about 0.5 °C. Under the RCP8.5 scenario, global warming would continue throughout the current century, and the global mean temperature would increase by almost 4 °C within 100 years. Under RCP4.5, the corresponding increase would be about 2 °C, whilst under RCP2.6, it would be slightly more than 1 °C. Considering the global warming already taking place before the baseline period of 1971–2000, the last scenario corresponds to a temperature increase of slightly less than 2 °C compared to the pre-industrial level. Regardless of future reductions in emissions, global warming will continue during the next few decades.

Compared to other regions on Earth, very intense warming has been simulated for northern polar areas in winter as a result of the partial disappearance of sea-ice cover. Conversely, in the northern Atlantic Ocean south of Iceland, warming will be modest because a weakening of the warm ocean current (the Gulf Stream) will

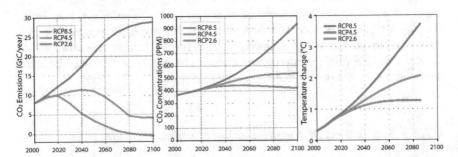

Fig. 3.1 Temporal evolution of global emissions in gigatonnes of carbon per year (left panel), atmospheric concentrations in parts per million by volume of CO_2 (central panel), and projected changes in global mean annual temperature in degrees Centigrade (right panel) for the period 2000–2085 under the RCP2.6, RCP4.5 and RCP8.5 scenarios. Temperature change is expressed relative to 1971–2000 and corresponds to the mean of simulations made using 28 different climate models (Ruosteenoja et al. 2016a; Venäläinen et al. 2020)

partially cancel the influence of global warming. Precipitation is projected to increase in equatorial areas. In addition, winter precipitation will increase at high latitudes. Decreasing precipitation totals are expected in multiple subtropical areas.

Globally, climate change will have multiple serious implications. In particular, the RCP8.5 scenario would lead to very severe climate change, with the consequences for many underdeveloped countries being catastrophic because agricultural production would suffer immensely from high temperatures and water shortages. This could lead to massive migrations from the developing world into wealthier countries. The rates of thermal expansion in ocean waters and the melting of continental glaciers and polar ice sheets would increase. The resulting sea-level rise would threaten numerous large coastal settlements and, consequently, a large proportion of Earth's population. In addition, the rapid environmental change would also threaten to drive a substantial share of the planet's plant and animal species to extinction.

Box 3.2 Representative Concentration Pathways (RCPs)

Future emissions, and the resulting atmospheric concentrations of the various greenhouse gases, cannot be known in advance, and therefore several alternative greenhouse gas scenarios have been developed. The emissions depend on the growth of the world population, energy consumption, energy production technologies, land use, etc. Since the IPCC's Fifth Assessment Report (AR5), future greenhouse gas scenarios called Representative Concentration Pathways (RCPs) have been used. The RCP2.6 scenario represents very low emissions, whilst RCP4.5 and RCP8.5 involve moderate and very high emissions, respectively. The number after the acronym refers to radiative forcing; that is, the imbalance between the solar radiation absorbed and the thermal infrared radiation emitted by the earth. For example, if the RCP4.5 scenario is realised, the positive (= warming) globally averaged radiative forcing at the end of the twenty-first century will be 4.5 Wm^{-2}. The RCP scenarios take account of the future emissions and atmospheric concentrations of several other greenhouse gases besides CO_2, as well as the aerosol particles. According to the RCP8.5 scenario, emissions would continue to increase throughout the twenty-first century, ultimately reaching three times the amount in 2000. The concentration of CO_2 would then approach 1000 ppm by 2100 (Fig. 3.1). In the RCP4.5 scenario, the CO_2 concentration would stabilise at close to 540 ppm. This level is about double that of the pre-industrial era. Under the most environmentally friendly RCP2.6 scenario, the concentrations would start to decrease slowly after the middle of this century. The RCP2.6 scenario would roughly meet the targets of the Paris Agreement. More information about the RCP scenarios is available in van Vuuren et al. (2011).

If the RCP2.6 scenario is realised, the consequences of the change will be far less severe than those resulting from RCP8.5. However, this target would require efficient reductions in global emissions starting right now, in the 2020s. This seems to be a huge challenge at present. Apart from the reduction in emissions, land-use changes, such as increasing or decreasing the share of forests, can impact greenhouse gas concentrations either adversely or favourably. Growing forests effectively absorbs CO_2 from the atmosphere. However, they also impact surface albedo, i.e. increasing albedo cooling the climate and decreasing albedo adding to the warming, respectively.

3.1.3   Projected Climate Change in Europe

In Europe, by the middle of this century, under RCP4.5, climate model simulations have projected the largest annual mean temperature increase — about 3 °C relative to the end of the nineteenth century — for the north-eastern part of the continent (Fig. 3.2). In Western Europe and along the coasts of the Mediterranean Sea, the projected warming is close to 2 °C. During the winter months, the warming in Northern Europe will be stronger than during the summer. Annual precipitation is projected to decrease in Southern Europe and increase in Northern and Central Europe. The maximum local annual changes would be around ±10%. The increase in precipitation in Northern Europe is projected to be the greatest in winter, whilst the decrease in precipitation will be the greatest in Southern Europe during summer.

The annual amount of solar radiation will increase across most of Europe. The largest increase — about 6% — is projected for Central Europe. Relative humidity is projected to decrease by 1–3 percentages. Temporal fluctuations in temperature will attenuate in the cold season, whereas fluctuations in precipitation will be amplified. Changes in their variability are expected to be strongest in the northern and north-eastern parts of the continent.

Under RCP4.5, the thermal growing season (defined as the period when daily mean temperatures are above 5 °C) is projected to lengthen by 10–15 days, both in the spring and autumn, from 1971–2000 to 2040—2069. Moreover, the temperature sum of the growing season is projected to increase by several hundreds of degree days. For example, around 2050, the average sum of growing degree days (GDDs, with base temperature of 5 °C) in southern Fennoscandia would be approximately the same as in northern Central Europe in the late twentieth century (Fig. 3.3).

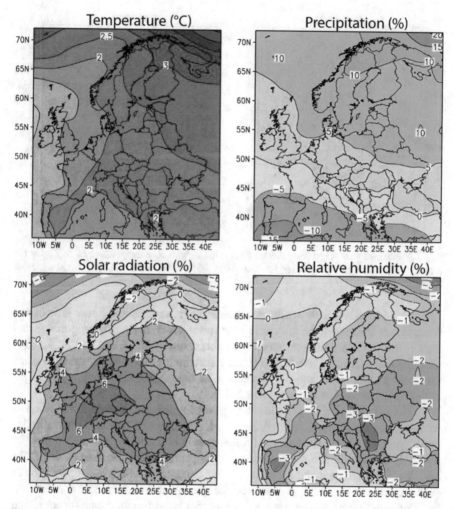

Fig. 3.2 Projected changes in annual mean temperature (°C), precipitation (%), incident solar radiation (%) and relative humidity (in percentage points) in Europe from the period 1971–2000 to 2040–2069 under RCP4.5 (Venäläinen et al. 2019)

3.2   Climate Change Impacts on Forests and Forestry

3.2.1   Forest Growth and Dynamics

Climate change is already having both direct and indirect effects on forests and forestry in different European regions. The direct effects include changes in the growing conditions of the forests due to changing temperatures, precipitation and atmospheric CO_2 concentrations. Indirect effects consist of various abiotic and biotic disturbances. In addition, land-use policy aimed at mitigating climate change

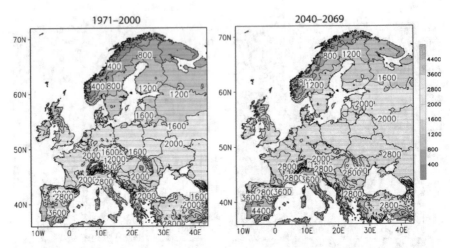

Fig. 3.3 Average sum of growing degree days (GDDs, with base temperature of 5 °C) for the period 1971–2000 and projection for the years 2040–2069 under RCP4.5 (redrawn from Ruosteenoja et al. 2016b)

can affect forests. As well as climate change and its severity, the future growth and dynamics of forests will be affected by the forest structure (i.e. the proportions and ages of tree species) and the intensity of forest management and harvesting (e.g. Heinonen et al. 2018). Climate change may have both positive and negative impacts on forest growth, such impacts depending on the geographical region and forest zone (European Environment Agency 2017).

In Northern Europe, longer and warmer growing seasons in general will promote more optimal forest growing conditions, especially in boreal forests at high latitudes and altitudes. This is because boreal forest growth is currently primarily limited by relatively short summers and low summer temperatures (Hyvönen et al. 2007). In addition, the increase in atmospheric CO_2 concentrations will favour forest growth (e.g. Hyvönen et al. 2007). In Southern Europe, but also to some extent in Central and Northern Europe, the growing conditions may become suboptimal for some tree species (Allen et al. 2010; Reyer et al. 2014). This is related to too high temperatures and too low soil water availability during the growing seasons. As a result, the growth of some tree species may slow down and mortality may increase. The differences in the responses among various tree species may be expected to increase in tandem with the severity of the climate change. In addition, the expected increase in many abiotic and biotic disturbances may counteract the positive effects of climate change on forest productivity, at least partially (Jactel et al. 2011; Reyer et al. 2017; Seidl et al. 2017).

3.2.2 Abiotic Disturbances

3.2.2.1 Wind and Snow Damage

During the last few decades, the major causes of widespread forest damage in Europe have been windstorms and forest fires (Schelhaas et al. 2003; Senf and Seidl 2021). In the period 1986–2016, storms were a major disturbance agent in Western and Central Europe, accounting locally for >50% of all disturbances (Senf and Seidl 2021). However, storm-related disturbances have also occurred in south-eastern and Eastern Europe. Fires have been a major disturbance agent in Southern and South-eastern Europe, but they have also occurred in Eastern and Northern Europe.

Strong winds have destroyed a significant amount of timber, causing substantial economic losses for forestry, especially in Central and Northern Europe. The increased amount of wind damage in the last few decades can be explained, at least partly, by an increasing volume of growing stock and changes in forest structures (Schelhaas et al. 2003). In Northern Europe, wind damage is likely to increase in the future because climate warming will shorten the duration of soil frost, which currently provides additional anchorage for trees during the windiest season of the year, from late autumn to early spring (Lehtonen et al. 2019). In addition, soil moisture is projected to increase in late autumn, likewise making forests more vulnerable to windfall.

According to multi-model-derived projections for European wind climate, climate change will not significantly alter the wind speeds in Northern Europe (Ruosteenoja et al. 2019). There is no robust signal of increasing or decreasing storminess in other European regions, either (e.g. Kjellström et al. 2018; Ruosteenoja et al. 2019). However, the projections for future trends in storminess diverge among the climate models (e.g. Feser et al. 2015). Accordingly, possible regional increases in the intensity of strong storms, changes in storm tracks, increasing growing stock and changes in forest structures (age and tree species composition) may affect the wind damage risks to forests.

Compared to the damage caused by windstorms, snow-induced damage in European forests is typically far less severe (Schelhaas et al. 2003). Snow-induced damage occurs most frequently in Northern Europe and at high altitudes (Nykänen et al. 1997). For most of Europe, climate model projections for the mid-twenty-first century indicate slightly decreasing probabilities for heavy snow loading. In northern Fennoscandia (e.g. northern and eastern Finland and north-western Russia), however, the probability of heavy snow loads may increase slightly (Groenemeijer et al. 2016; Lehtonen et al. 2016a). Excessive snow loads typically result in stem breakage and the bending or leaning of tree stems. In particular, young Scots pines (*Pinus sylvestris*) and broadleaf trees with a large height-to-stem-diameter ratio are susceptible to snow damage (Nykänen et al. 1997). With unfrozen soil, trees can be uprooted. The increase in duration of frost-free periods is expected to increase such damage under the warming climate.

In addition to climatic factors, the severity of wind and snow damage risk is affected by the tree and stand characteristics (tree species, height and diameter, rooting characteristics, and stand density) and the forest configuration (e.g. the distance from the upwind edge of a new clearcut). For example, in high-risk areas of the boreal zone, an increase in the cultivation of the shallow-rooted Norway spruce (*Picea abies*) at the cost of Scots pine will increase the future wind damage risk (Ikonen et al. 2020). Conversely, an increase in the cultivation of pine and broadleaf trees will increase the future snow damage risk (Nykänen et al. 1997). Trees damaged by wind or snow may also bend over or lean on power lines, and thus may disrupt the availability of electricity to society.

3.2.2.2 Drought and Forest Fires

Global climate change is expected to increase the occurrence of summer drought everywhere in Europe, most severely in the south, but to some extent in the north as well. This increasing drought will be caused by an intensification in potential evaporation, which will outweigh the impact of changes in precipitation. In Northern Europe, the average moisture in the soil surface layer will decrease, especially in spring and early summer, whereas in Southern Europe, the loss will be most pronounced in late summer (Fig. 3.4). Consequently, anomalously dry conditions are projected to become increasingly frequent in European forests (Ruosteenoja et al. 2018). Accordingly, at sites with water shortages, in particular, forest growth is

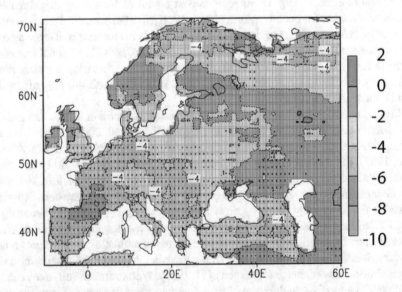

Fig. 3.4 Projected changes in time-mean near-surface soil moisture (in percentage points) in Europe in June–August under RCP4.5 for the period 2040–2069. The change was averaged over 26 GCMs and is expressed relative to the period 1971–2000. Areas where at least 23 models agreed on the sign of change are stippled (Ruosteenoja et al. 2018)

expected to decline and mortality to increase under a warmer climate (Allen et al. 2010).

High temperatures and an increase in the frequency and severity of summer drought periods will act to increase the risk of forest fires. This phenomenon has already been observed, particularly in South-eastern Europe (Venäläinen et al. 2014). The widespread, devastating fires in Sweden in the summers of 2014 and 2018 showed that large-scale forest fires are possible in the Nordic countries as well. For example, a single fire in Sweden in 2014 burned 14,000 ha of forest (Joint Research Centre 2015).

In the southern parts of Europe, the meteorological fire danger is projected to increase significantly by the middle of this century (e.g. Groenemeijer et al. 2016). It is likely that the fire danger will likewise increase in Northern Europe (Lehtonen et al. 2016b). In semi-arid areas, such as the Mediterranean region, low vegetation productivity may limit these fires, and therefore the actual occurrence of fires may increase less drastically than what is predicted by the changing weather conditions alone. However, even when considering ecosystem functioning, the area burned is still likely to increase, especially in the Mediterranean Basin, the Balkan region and Eastern Europe (Migliavacca et al. 2013; Turco et al. 2018). Under a warmer and dryer climate, there may be an increasing risk for mega-scale forest fires in European forests, such as those that have recently occurred in Canada and Siberia (e.g. Hanes et al. 2019; Walker et al. 2019). Such disturbances could release huge amounts of stored carbon into the atmosphere, thus nullifying the potential positive impact on climate change on carbon sequestration in forests.

3.2.3   Biotic Disturbances

3.2.3.1   European Spruce Bark Beetle Outbreaks

In recent decades, disturbances from bark beetles have greatly increased in Europe. The amount of timber damaged by bark beetles in spruce and pine forest has increased by nearly 70% over the last 40 years, from 2.2 million m^3 per year (1971–1980) to 14.5 million m^3 per year (2002–2010) (Seidl et al. 2014). The planting of Norway spruce outside of its natural range (and on sites with lower soil water holding capacity), an increase in growing stocks, and changes in forest age structures and compositions have made forests more prone to bark beetle outbreaks (Hlásny et al. 2019; Jandl 2020). In addition to warm and dry summer conditions, severe wind damage and drought also intensify bark beetle outbreaks (Marini et al. 2017).

The primary bark beetle species in Europe responsible for outbreaks is the widely distributed, eight-toothed European spruce bark beetle (*Ips typographus*) (e.g. Christiansen and Bakke 1988). At low population levels, it colonises only stressed and dying trees (e.g. wind-damaged Norway spruce). However, at high population levels, it can mount a mass attack on a large number of healthy trees. European spruce bark beetle particularly favours older and larger trees (e.g. aged >60 years, diameter at breast height > 20–25 cm) (Hlásny et al. 2019). European spruce bark beetle outbreaks have largely increased in recent years in Europe (Hlásny et al. 2019, 2021; Jandl 2020; Romashkin et al. 2020). For example, in Czechia in 2017, in an unforeseen, severe outbreak, the amount of damaged timber exceeded the annual demand at the country level, collapsing the timber market and prices, respectively. In Austria over the last decade, the high supply of beetle-infested timber has reduced the market price for bark beetle affected timber to 30% of the previous level (Jandl 2020). In Sweden, a European spruce bark beetle outbreak damaged an additional 4 million m^3 of timber after windstorm Gudrun, which damaged 70 million m^3 of timber in January 2005 (Lindelöw and Schroeder 2008).

The survival and reproduction of European spruce bark beetle benefit from warmer and dryer climates, and thus also from climate warming (Christiansen and Bakke 1988; Jönsson et al. 2007; Lindelöw and Schroeder 2008; Hlásny et al. 2019; Jandl 2020). Under optimal conditions, bark beetle populations can increase more than 15-fold from one generation to the next (Hlásny et al. 2019). It can also produce two generations (multivoltinism) in one summer, if swarming conditions are favourable early in the season, and the sum of GDDs exceeds approximately 1500 °C days, which is twice the GDD sum needed for the complete development of an individual, from egg to adult (625–750 GDDs) (Jönsson et al. 2007). Moreover, the number of successfully developed beetles in different sister broods of the first generation increase with an increase in GDD sum (Öhrn et al. 2014). In warm areas of the southern part of the species distribution region, a third generation may also be possible (Jakoby et al. 2019).

Lower GDDs currently partially explain the lower bark beetle outbreak risk in Northern Europe compared with more southerly areas. However, the 1500 GDD isoline that potentially allows a change from univoltine (i.e. a single generation in summer) to multivoltine population dynamics is moving northwards. For example, in European Russia, the latitudinal shift of the isoline that indicates the northern limit of 1500 GDD has moved 450 km northwards since the 1960s (Romashkin et al. 2020).

Based on Asikainen et al. (2019), the probability of exceeding the GDD sum of a 1500 °C-day threshold will increase in Northern Europe under climate change. Recent warmer and drier summers, together with unharvested wood left in forests after wind damage, have already increased the populations and attacks of bark beetles in the southern boreal zone, and even in middle boreal zone (Romashkin et al. 2020). Overall, European spruce bark beetle outbreaks are projected to increase in the future, under warmer and drier climates, from Central to Northern Europe (Jönssön et al. 2007; Seidl et al. 2014).

3.2.3.2 Other Biotic Threats to Forest Health

European spruce bark beetle is currently the most obvious biotic damage agent in European forests, outbreaks of which have markedly increased with climate warming. However, climate change also affects the reproduction, growth, behaviour and potential distribution range of other species that can cause problems with forest health. Thus, disturbances by several other major forest pathogens, pest insects and browsing mammal species are also expected to increase in European forests.

In Northern Europe, *Heterobasidion* spp. root rot is already one of the most destructive diseases in conifers (Garbelotto and Gonthier 2013). However, increasing temperatures are further increasing its spore formation and the growth rate of its mycelia. Milder winters increase the length of the period the fungus is able to spread and infect new stands (La Porta et al. 2008). Together, these intensify the amount of decay in infected trees and the spread of fungus in diseased stands.

The epiphytic, parasitic vascular plant, pine mistletoe (*Viscum album* ssp. *austriacum*), is also increasing in abundance at its current northern limit, such as in Germany and Poland, and is spreading upwards into the montane forests of Europe (Szmidla et al. 2019). This is probably the result of increasing winter temperatures in areas where pine mistletoe has previously been limited by the low freeze tolerance of its seeds. Abundant mistletoe populations reduce tree growth substantially and, in dry areas, they also increase water stress and tree mortality (Kollas et al. 2018).

Higher temperatures are likely to promote distributional shifts in many native forest pest-insect species and invasive alien species towards more northerly latitudes and higher elevations (Battisti and Larsson 2015). Frequent cold winters in Northern Europe have so far limited outbreaks of many insect defoliators that overwinter as eggs. However, an increase in winter temperatures will favour their reproduction and, concurrently, may increase the risk from these in the future. Nun moth (*Lymantria monacha*), one of the most serious defoliators of coniferous forests in Central Europe (Bejer 1988), is a good example. Previously cold winters have controlled nun moth populations in Northern Europe because its eggs freeze in temperatures below −30 °C (Fält-Nardmann et al. 2018). The species has been historically absent or very rare in Finland, but since the 1990s, its populations have increased hugely, and it is now very abundant in the southern part of the country (Melin et al. 2020).

In Southern Europe, heat-tolerant and cold-sensitive species, such as the pine processionary moth (*Thaumetopoea pityocampa*) and the oak processionary moth (*Thaumetopoea processionea*) have expanded their geographical ranges beyond the Mediterranean region (Battisti et al. 2005; Godefroid et al. 2020). Processionary moths damage not only trees, but their larvae have defensive hairs (urticating setae that the larvae release when disturbed) that can cause allergic reactions in humans (Vega et al. 2011). Therefore, the processionary moth is considered to be a threat to human health when present in urban forests and parks (Rossi et al. 2016).

The warming climate is increasing problems relating to the regeneration of coniferous forests in Europe by, for example, the large pine weevil (*Hylobius abietis*) (Nordlander et al. 2017). This is because warmer summers and a shortening of the frozen soil period is decreasing the development time of immature weevils, increasing their feeding time and prolonging the feeding period. Browsing by high local populations of moose (*Alces alces*) is also a serious problem in young Scots pine and birch seedling stands in Northern Europe. The expected reduction in snow depth and duration may increase the severity of browsing damage (e.g. Herfindal et al. 2015).

3.3 Climate Change, Adaptation and Risk Management

Forests should provide multiple ecosystem services for society. However, climate change is inducing many abiotic and biotic damage risks in forests and forestry at different spatial and temporal scales, all of which affect the provisioning of ecosystem services. Warmer and drier summer conditions particularly increase the risk of damage by drought, forest fires and pest insects, while warmer and wetter winters increase the risk of damage by windstorms and strong winds, heavy snow loading and pathogens (Seidl et al. 2017). Such disturbances are likely to increase the most in coniferous forests in the boreal zone. They may partially counteract the positive effects of climate change on forest productivity, causing severe economic losses in forests (Hanewinkel et al. 2013; Reyer et al. 2017).

The simultaneous occurrence of multiple hazardous events can make the adverse impacts manifold (Hanewinkel et al. 2013; Venäläinen et al. 2020; Hlásny et al. 2019). Wind and snow damage in particular, but also the occurrence of drought, may increase the availability of breeding material for bark beetles, thus enhancing their outbreaks. The drought may further influence the forest fire risk through increased tree mortality (e.g. Jenkins et al. 2014). Wind and snow damage may also increase *Heterobasidion* spp. attacks through tree injuries from harvesting, which will then exacerbate the risk of wind damage due to poorer anchorage and less stem resistance in decaying-wood trees.

How vulnerable forests are to climate change and the associated increase in various abiotic and biotic disturbances depends on the exposure (e.g. the severity of the climate change, the climate variability and its extremes), sensitivity and adaptive capacity of forests. Fortunately, adaptive forest management can offer ways to increase the resilience of forests to climate change and its related disturbances. The severity of climate change will affect the necessary adaptation and risk management actions for different regions and time spans. In adaptation and risk management, the occurrence of multiple hazardous events should be considered simultaneously in order to ensure the sustainable provisioning of different ecosystem services for society. Fortunately, the same management measures may simultaneously enhance the resilience of forests against multiple abiotic and biotic disturbances (Table 3.1).

The resilience of forests against different abiotic and biotic disturbances may be increased, for example, by modifying the age structure and tree species composition at the forest landscape level through forest management. In forest regeneration, the

Table 3.1 Possible adaptive and risk management strategies

Possible management strategies for enhancing resilience

High temperature/drought
- region–/site-specific species/genotype choice
- natural regeneration where appropriate
- mixed conifer–deciduous stands
- wider spacing and heavier thinning regimes
- shorter rotation periods (or lower target diameters for final harvesting)

Wind damage
- region–/site-specific species choice
- timely pre-commercial and commercial thinning (not too heavy)
- avoidance of forest fertilisation at the same time as thinning
- avoidance of heavy thinning in the upwind edges of new openings
- avoidance of creating large height differences between adjacent stands in final harvesting

Snow damage
- region–/site-specific species choice
- timely pre-commercial and commercial thinning (not too heavy in dense stands)
- avoidance of forest fertilisation on sites at high altitudes (>200 m a.s.l.)

Bark beetle outbreak
- mixed conifer–deciduous stands
- timely thinning to improve tree vigour (outbreak prevention)
- shorter rotation periods (or lower target diameters)
- harvesting of infested trees (sanitation felling and salvage logging)
- removal of harvested and wind-damaged trees before beetles fly in spring/emergence of first new beetle generation
- mosaic of forest stands in forest landscapes to minimise spread of beetles

Heterobasidion **root rot**
- mixed conifer–deciduous stands
- shorter rotation periods (or lower target diameters for final harvesting)
- harvesting of unhealthy trees

Forest fires
- fragmented forest landscape to limit fire spread
- timely thinning to avoid mortality (decrease in flammable material)

appropriate region- and site-specific choice of tree species (genotypes) and spacing may increase the adaptive capacity and resilience of the forest in the long term. Similarly, favouring more resilient tree species in pre-commercial (tending) and commercial thinning may increase the resilience of the forest. By favouring mixtures of conifers and broadleaf species over monocultures on suitable sites, their resilience may be further increased against many abiotic and biotic risks to forests (e.g. Pretzsh et al. 2017). For example, the wind damage risk in forests with shallow-rooting Norway spruce are well known throughout Europe (Jandl 2020). Overall, single-species forests offer pests and pathogens more opportunities for spreading than mixed stands, where tree species have different ecological niches. The latter scenario provides, for example, fewer host trees for a bark beetle outbreak and could also host larger populations of their natural enemies and competitors, etc. (Hlásny et al. 2019). The use of greater thinning intensity or wider spacing increases water availability at the tree level in a stand, which may decrease drought stress in trees, and its consequent damage.

The avoidance of fertilisation in high-altitude forest sites, especially in relation to thinning, may decrease the snow damage risk to boreal forests (e.g. Valinger and Lundqvist 1992; Nykänen et al. 1997). Furthermore, the use of shorter rotation periods or lower target diameters for final harvesting may decrease the risk of damage by windstorms and strong winds, pest insects (e.g. bark beetles) and pathogens (e.g. wood decay by *Heterobasidion*), for example, in Norway spruce, which is particularly sensitive to such damage. The increase in risk of large-scale forest fires during summer droughts (Ruosteenoja et al. 2018) must also be considered in the timing of forest harvesting operations because the sparks generated by the machinery used in such activities may result in the ignition of forest fires.

Uncertainties relating to climate change, forest disturbances and the future preferences of society call for the simultaneous use of diverse management strategies, rather than a single, one-size-fits-all management strategy (e.g. even-, uneven- and any-aged management), which might also help to increase the overall production levels of ecosystem services (Díaz-Yáñez et al. 2020). Multi-functionality in forest management may also ensure the simultaneous provisioning of different ecosystem services for society, whilst increasing the resilience of forests against abiotic and biotic disturbances. However, the frequent adjustment of forest management practices (e.g. 10–20-year frequency) to changing growing conditions is also needed in order to adapt to climate change and maintain forest resilience, which are required to sustain the provisioning of different ecosystem services. On the other hand, climate change may increase large-scale forest fire and pest insect occurrences in unmanaged, mature forests (e.g. in forest conservation areas) due to the increased tree mortality that will result from warmer and drier climates. As a result, these disturbances may also spread to managed forests. Thus, preparedness for such risks should be increased in society.

Overall, the challenge of dealing with climate change-induced disturbances in forest management and forestry is pan-European (Jandl 2020). Different adaptation and risk management actions may be needed, depending on geographical region and time span, to maintain the sustainable provisioning of different ecosystem services for society, and to increase the forest resilience. The role of forests in climate change mitigation should also be considered in adaptation and risk management. This is because forests contribute greatly to climate change mitigation through sequestering carbon from the atmosphere and storing it in forest ecosystems and wood-based products, the latter also substituting for fossil-intensive resources (Kauppi et al. 2018). The intensity of forest management practices and the severity of natural disturbances may significantly affect the carbon sequestration (and stock) in forests as a result of changes in forest structure (e.g. age and tree species composition). Consequently, changes in forest structure will indirectly affect climate regulation through changes in forest albedo and latent heat fluxes, biogenic volatile organic compounds and aerosols (e.g. Thom et al. 2017).

3.4 Research Implications

There are large uncertainties in predicting future climate and its impacts on European forests and forestry. This is due to uncertainties in global developments in future greenhouse gas emissions, which are greatly affected by the level of success of climate change mitigation. Therefore, such uncertainties should be considered in climate change impact and adaptation studies, by using several alternative climate projections in simulation-based scenario analyses. In order to define climate-smart (and adaptive) risk management strategies, there is a need for a more holistic understanding of how the prevailing climatic conditions, forest structure, forest management (strategies) and severity of climate change, together with the associated increases in natural disturbances, may affect the provisioning of multiple ecosystem services (e.g. timber, biodiversity and the recreational values of forests) and climate regulation for different geographical regions and time spans. Climate change will affect, in addition to the physiological conditions of trees and tree defence mechanisms against natural enemies, the distribution and population dynamics of those enemies, and this needs to be understood in greater detail. In the current world of uncertainty, we should seek different ways to simultaneously improve the provisioning of different ecosystem services for society, the resilience of forests and their climate benefits.

3.5 Key Messages

- There are large uncertainties in the projected climate change and its impacts on European forests and forestry for different regions and time spans, due to large uncertainties in the level of success of climate change mitigation efforts.
- In general, forest growth is projected to increase in Northern Europe, as opposed to Southern Europe.
- Climate change may induce multiple abiotic and biotic damage risks in forests and forestry throughout Europe via windstorms, drought, forest fires, bark beetle outbreaks and wood-decaying fungus diseases.
- The uncertainties relating to climate change and the increasing multiple risks to forests and forestry should be considered when adapting forest management and forestry to climate change in order to increase the resilience of forests against different abiotic and biotic disturbances.
- The necessary adaptation and risk management measures may differ, depending on geographical region and time span.

Acknowledgments Heli Peltola thanks the financial support from the OPTIMAM project (grant number 317741) funded by the Academy of Finland.

References

Allen CD, Macalady AK, Chenchouni H, Bachelet D, McDowell N, Vennetier M et al (2010) A global overview of drought and heat-induced tree mortality reveals emerging climate change risks for forests. For Ecol Manag 259:660–684. https://doi.org/10.1016/j.foreco.2009.09.001

Asikainen A, Viiri H, Neuvonen S, Nevalainen S, Lintunen J, Laturi J, … Ruosteenoja K (2019) Ilmastonmuutos ja metsätuhot – Analyysi ilmaston lämpenemisen seurauksista Suomen osalta [Climate change and forest damage in Finland]. The Finnish climate change panel, Helsinki, Finland. https://www.ilmastopaneeli.fi/wp-content/up-loads/2019/01/Ilmastopaneeli_Mets%C3%A4tuhoraportti_tiivistelm%C3%A4-1.pdf. Accessed 25 Jan 2021

Battisti A, Larsson S (2015) Climate change and insect pest distribution range. In: Björkman C, Niemelä P (eds) Climate change and insect pests, Climate Change Series. CABI, Wallingford, pp 1–15. https://doi.org/10.1079/9781780643786.0001

Battisti A, Stastny M, Netherer S, Robinet C, Schopf A, Roques A, Larsson S (2005) Expansion of geographic range in the pine processionary moth caused by increased winter temperatures. Ecol Appl 15:2084–2096. https://doi.org/10.1890/04-1903

Bejer B (1988) The nun moth in European forests. In: Berryman AA (ed) Dynamics of forest insect populations. Springer, Boston, pp 211–231

Christiansen E, Bakke A (1988) The spruce bark beetle of Eurasia. In: Berryman AA (ed) Dynamics of forest insect populations, patterns, causes, and implications. Plenum Press, New York, pp 480–505

Díaz-Yáñez O, Pukkala T, Packalen P, Peltola H (2020) Multifunctional comparison of different management strategies in boreal forests. For An Int J For Res 93(1):84–95. https://doi.org/10.1093/for-estry/cpz053

European Environment Agency (2017) Climate change, impacts and vulnerability in Europe 2016. An indicator based report. European Environment Agency Report 1/2017. https://www.eea.europa.eu/publications/climate-change-im-pacts-and-vulnerability-2016. Accessed 25 Jan 2021

Fält-Nardmann JJ, Ruohomäki K, Tikkanen OP, Neuvonen S (2018) Cold hardiness of Lymantria monacha and L. dispar (Lepidoptera: Erebidae) eggs to extreme winter temperatures: implications for predicting climate change impacts. Ecol Entomol 43(4):422–430. https://doi.org/10.1111/een.12515

Feser F, Barcikowska M, Krueger O, Schenk F, Weissea R, Xiae L (2015) Review article: storminess over the North Atlantic and northwestern Europe – a review. Q J R Meteorol Soc 141:350–382. https://doi.org/10.1002/qj.2364

Garbelotto M, Gonthier P (2013) Biology, epidemiology, and control of Heterobasidion species worldwide. Annu Rev Phytopathol 51:39–59. https://doi.org/10.1146/annurev-phyto-082712-102225

Godefroid M, Meurisse N, Groenen F, Kerdelhué C, Rossi JP (2020) Current and future distribution of the invasive oak processionary moth. Biol Invasions 22(2):523–534. https://doi.org/10.1007/s10530-019-02108-4

Groenemeijer P, Vajda A, Lehtonen I, Kämäräinen M, Venäläinen A, Gregow H., … Púčik T (2016) Present and future probability of meteorological and hydrological hazards in Europe. Technical Report Report number: 608166-D 2.5, 165 pp. Retrieved from http://resolver.tudelft.nl/uuid:906c812d-bb49-408a-aeed-f1a900ad8725

Hanes C, Wang X, Jain P, Parisien M-A, Little J, Flannigan M (2019) Fire-regime changes in Canada over the last half century. Can J For Res 49(3):256–269. https://doi.org/10.1139/cjfr-2018-0293

Hanewinkel M, Cullmann DA, Schelhaas MJ, Nabuurs GJ, Zimmermann NE (2013) Climate change may cause severe loss in the economic value of European forestland. Nat Clim Chang 3:203. https://doi.org/10.1038/NCLIMATE1687

Heinonen T, Pukkala T, Kellomäki S, Strandman H, Asikainen A, Venäläinen A, Peltola H (2018) Effects of forest man- agement and harvesting intensity on the timber supply from Finnish forests in a changing climate. Can J For Res 48(10):1124–1134. https://doi.org/10.1139/cjfr-2018-0118

Herfindal I, Tremblay JP, Hester AJ, Lande US, Wam HK (2015) Associational relationships at multiple spatial scales affect forest damage by moose. For Ecol Manag 348:97–107. https://doi.org/10.1016/j.foreco.2015.03.045

Hlásny T, Krokene P, Liebhold A, Montagné-Huck C, Müller J, Qin H., … Viiri H (2019) Living with bark beetles: impacts, outlook and management options. From science to policy 8. European Forest Institute https://efi.int/sites/default/files/files/publication-bank/2019/efi_fstp_8_2019.pdf. Accessed 25 Jan 2021

Hlásny T, Zimova S, Merganicova K, Stepanek P, Modlinger R, Turcani M (2021) Devastating outbreak of bark beetles in the Czech Republic: drivers, impacts, and management implications. For Ecol Manag 490:119075. https://doi.org/10.1016/j.foreco.2021.119075

Hyvönen R, Ågren GI, Linder S, Persson T, Cotrufo MF, Ekblad A et al (2007) The likely impact of elevated [CO2], nitrogen deposition, increased temperature and management on carbon sequestration in temperate and boreal forest ecosystems: a literature review. New Phytol 173:463–480. https://doi.org/10.1111/j.1469-8137.2007.01967.x

Ikonen V-P, Kilpeläinen A, Strandman H, Asikainen A, Venäläinen A, Peltola H (2020) Effects of using certain tree species in forest regeneration on regional wind damage risks in Finnish boreal forests under different CMIP5 projections. Eur J For Res 139:685–707. https://doi.org/10.1007/s10342-020-01276-6

Jactel H, Petit J, Desprez-Loustau M-L, Delzon S, Piou D, Battisti A, Koricheva J (2011) Drought effects on damage by forest insects and pathogens: a meta-analysis. Glob Change Biol 18(1). https://doi.org/10.1111/j.1365-2486.2011.02512.x

Jakoby O, Lischke H, Wermelinger B (2019) Climate change alters elevational phenology patterns of the European spruce bark beetle (Ips typographus). Glob Change Biol 25(12):4048–4063. https://doi.org/10.1111/gcb.14766

Jandl R (2020) Climate-induced challenges of Norway spruce in northern Austria. Trees For People 1:100008. https://doi.org/10.1016/j.tfp.2020.100008

Jenkins M, Runyon J, Fettig C, Page W, Bentz B (2014) Interactions among the mountain pine beetle, fires, and fuels. For Sci 60(3):489–501. https://doi.org/10.5849/forsci.13-017

Joint Research Centre (2015) Forest fires in Europe, Middle East and North Africa 2014. JRC Technical Reports. Joint report of JRC and Directorate-General Environment. https://doi.org/10.2788/224527

Jönsson AM, Harding S, Bärring L, Ravn HP (2007) Impact of climate change on the population dynamics of Ips typographus in southern Sweden. Agric For Meteorol 146:70–81. https://doi.org/10.1016/j.agrformet.2007.05.006

Kauppi P, Hanewinkel M, Lundmark T, Nabuurs G-J, Peltola H, Trasobares A, Hetemäki L (2018) Climate smart for- estry in Europe. European Forest Institute. https://efi.int/sites/default/files/files/publication-bank/2018/Cli-mate_Smart_Forestry_in_Europe.pdf. Accessed 25 Jan 2021

Kjellström E, Nikulin G, Strandberg G, Christensen O, Jacob D, Keuler K et al (2018) European climate change at global mean temperature increases of 1.5 and 2°C above pre-industrial conditions as simulated by the EURO- CORDEX regional climate models. Earth Syst Dynam 9:459–478. https://doi.org/10.5194/esd-9-459-2018

Kollas C, Gutsch M, Hommel R, Lasch-Born P, Suckow F (2018) Mistletoe-induced growth reductions at the forest stand scale. Tree Physiol 38:735–744. https://doi.org/10.1093/treephys/tpx150

La Porta N, Capretti P, Thomsen IM, Kasanen R, Hietala AM, Von Weissenberg K (2008) Forest pathogens with higher damage potential due to climate change in Europe. Can J Plant Pathol 30:177–195. https://doi.org/10.1080/07060661.2008.10540534

Lehtonen I, Kämäräinen M, Gregow H, Venäläinen A, Peltola H (2016a) Heavy snow loads in Finnish forests respond regionally asymmetrically to projected climate change. Nat Hazards Earth Syst Sci 16:2259–2271. https://doi.org/10.5194/nhess-16-2259-2016

Lehtonen I, Venäläinen A, Kämäräinen M, Peltola H, Gregow H (2016b) Risk of large-scale forest fires in boreal forests in Finland under changing climate. Nat Hazards Earth Syst Sci 16:239–253. https://doi.org/10.5194/nhess-16-239-2016

Lehtonen I, Venäläinen A, Kämäräinen M, Asikainen A, Laitila J, Anttila P, Peltola H (2019) Projected decrease in wintertime bearing capacity on different forest and soil types in Finland under a warming climate. Hydrol Earth Syst Sci 23:1611–1631. https://doi.org/10.5194/hess-23-1611-2019

Lindelöw A, Schroeder M (2008) The storm "Gudrun" and the spruce bark beetle in Sweden. Forstschutz 29 Aktuell 44:5–7. Retrieved from https://www.researchgate.net/publication/267208040.

Marini L, Økland B, Jönsson AM, Bentz B, Carroll A, Forster B et al (2017) Climate drivers of bark beetle outbreak dynamics in Norway spruce forests. Ecography 40:1426–1435. https://doi.org/10.1111/ecog.02769

Melin M, Viiri H, Tikkanen O-P, Elfving R, Neuvonen S (2020) From a rare inhabitant into a potential pest – status of the nun moth in Finland based on pheromone trapping. Silva Fenn 54(1):10262. https://doi.org/10.14214/sf.10262

Migliavacca M, Dosio A, Camia A, Hobourg R, Houston-Durrant T, Kaiser JW et al (2013) Modeling biomass burning and related carbon emissions during the 21st century in Europe. J Geophys Res Biogeosci 118:1732–1747. https://doi.org/10.1002/2013JG002444

Nordlander G, Mason EG, Hjelm K, Nordenhem H, Hellqvist C (2017) Influence of climate and forest management on damage risk by the pine weevil Hylobius abietis in northern Sweden. Silva Fenn 51(5). doi:10.14214/:sf.7751

Nykänen M-L, Peltola H, Quine CP, Kellomäki S, Broadgate M (1997) Factors affecting snow damage of trees with particular reference to European conditions. Silva Fenn 31(2):193–213

Öhrn P, Långström B, Lindelöw Å, Björklund N (2014) Seasonal flight patterns of Ips typographus in southern Sweden and thermal sums required for emergence. Agric For Entomol 16:1–23. https://doi.org/10.1111/afe.12044

Pretzsch H, Forrester DI, Bauhus J (2017) Mixed-species forests. Springer, Berlin

Reyer C, Lasch-Born P, Suckow F, Gutsch M, Murawski A, Pilz T (2014) Projections of regional changes in forest net primary productivity for different tree species in Europe driven by climate change and carbon dioxide. Ann For Sci 71(2):211–225. https://doi.org/10.1007/s13595-013-0306-8

Reyer C, Bathgate S, Blennow K, Borges JG, Bugmann H, Delzon S et al (2017) Are forest disturbances amplifying or cancelling out climate change-induced productivity changes in European forests? Environ Res Lett 12(3):034027. https://doi.org/10.1088/1748-9326/aa5ef1

Romashkin I, Neuvonen S, Tikkanen O-P (2020) Northward shift in temperature sum isoclines may favour Ips typogra- phus outbreaks in European Russia. Agric For Entomol 22:238–249. https://doi.org/10.1111/afe.12377

Rossi JP, Imbault V, Lamant T, Rousselet J (2016) A citywide survey of the pine processionary moth Thaumetopoea pityocampa spatial distribution in Orléans (France). Urban For Urban Green 20:71–80. https://doi.org/10.1016/J.UFUG.2016.07.015

Ruosteenoja K, Jylhä K, Kämäräinen M (2016a) Climate projections for Finland under the RCP forcing scenarios. Geo- physica 51(17–50) http://www.geophysica.fi/pdf/geophysica_2016_51_1-2_017_ruosteenoja.pdf. Accessed 25 Jan 2021

Ruosteenoja K, Räisänen J, Venäläinen A, Kämäräinen M (2016b) Projections for the duration and degree days of the thermal growing season in Europe derived from CMIP5 model output. Int J Climatol 36:3039–3055. https://doi.org/10.1002/joc.4535

Ruosteenoja K, Markkanen T, Venäläinen A, Räisänen P, Peltola H (2018) Seasonal soil moisture and drought occurrence in Europe in CMIP5 projections for the 21st century. Clim Dyn 50:1177–1192. https://doi.org/10.1007/s00382-017-3671-4

Ruosteenoja K, Vihma T, Venäläinen A (2019) Projected changes in European and North Atlantic seasonal wind climate derived from CMIP5 simulations. J Clim 32:6467–6490. https://doi.org/10.1175/JCLI-D-19-0023.1

Schelhaas M-J, Nabuurs G-J, Schuck A (2003) Natural disturbances in the European forests in the 19th and 20th centuries. Glob Change Biol 9:1620–1633. https://doi.
org/10.1046/j.1365-2486.2003.00684.x

Seidl R, Schelhaas M-J, Rammer W, Verkerk PJ (2014) Increasing forest disturbances in Europe
and their impact on carbon storage. Nat Clim Chang 4(9):806–810. https://doi.org/10.1038/
nclimate2318

Seidl R, Thom D, Kautz M, Martin-Benito D, Peltoniemi M, Vacchiano G et al (2017) Forest
disturbances under climate change. Nat Clim Chang 7:395–402. https://doi.org/10.1038/
nclimate3303

Senf C, Seidl R (2021) Storm and fire disturbances in Europe: distribution and trends. Glob Change
Biol. https://doi.org/10.1111/gcb.15679

Szmidla H, Tkaczyk M, Plewa R, Tarwacki G, Sierota Z (2019) Impact of common mistletoe
(Viscum album L.) on scots pine forests — a call for action. Forests 10(847). https://doi.
org/10.3390/f10100847

Thom D, Rammer W, Seidl R (2017) The impact of future forest dynamics on climate: interactive
effects of changing vegetation and disturbance regimes. Ecol Monogr 87:665–684. https://doi.
org/10.1002/ecm.1272

Turco M, Rosa-Cánovas JJ, Bedia J, Jerez S, Montávez JP, Llasat MC, Provenzale A (2018)
Exacerbated fires in Medi- terranean Europe due to anthropogenic warming projected
with non-stationary climate-fire models. Nat Commun 9:3821. https://doi.org/10.1038/
s41467-018-06358-z

Valinger E, Lundqvist L (1992) Influence of thinning and nitrogen fertilization on the frequency of
snow and wind in- duced stand damage in forests. Scott For 46:311–332

van Vuuren DP, Edmonds J, Kainuma M, Riahi K, Thomson A, Hibbard K et al (2011) The
representative con- centration pathways: an overview. Clim Chang 109:5–31. https://doi.
org/10.1007/s10584-011-0148-z

Vega J, Vega JM, Moneo I (2011) Skin reactions on exposure to the pine processionary caterpillar
(Thaumetopoea pityocampa). Actas Dermosifiliogr (English Ed) 102(9):658–667

Venäläinen A, Korhonen N, Koutsias N, Xystrakis F, Urbieta IR, Moreno JM (2014) Temporal
variations and change in forest fire danger in Europe for 1960–2012. Nat Hazards Earth Syst
Sci 14:1477–1490. https://doi.org/10.5194/nhess-14-1477-2014

Venäläinen A, Lehtonen I, Ruosteenoja K (2019) Projections of future climate for Europe, Uruguay
and China with implications on forestry. Finnish Meteorological Institute, Reports 2019:3.
https://doi.org/10.35614/isbn.9789523360853

Venäläinen A, Lehtonen I, Laapas M, Ruosteenoja K, Tikkanen O-P, Viiri H, Ikonen V-P, Peltola H
(2020) Climate change induces multiple risks to boreal forests and forestry in Finland: a literature review. Glob Change Biol 26:4178–4196. https://doi.org/10.1111/gcb.15183

Walker XJ, Baltzer JL, Cumming SG, Day NJ, Ebert C, Goetz S et al (2019) Increasing wildfires threaten historic carbon sink of boreal forest soils. Nature 572:520–523. https://doi.
org/10.1038/s41586-019-1474-y

Chapter 4
Outlook for the Forest-Based Bioeconomy

Elias Hurmekoski, Lauri Hetemäki, and Janne Jänis

Abstract The state of the world's managed forests is determined by the societal demands for wood resources and other ecosystem services. The forest-based sector is experiencing a number of structural changes, which makes the task of looking ahead important, but challenging. One of the main trends in the forest-based industries is diversification. On one hand, this refers to the emergence of new factors influencing the demand for forest-based products, which leads to substitution between forest-based products and alternative products. On the other hand, it refers to new market opportunities for forest-based industries in, for example, the construction, textiles, packaging, biochemicals and biofuels markets. As the importance of some of the traditional forest-based industries, such as communication papers, is declining, and new opportunities are simultaneously emerging, the sector will not necessarily be dominated by single sectors in the long term. However, research illuminating the possible impacts of the expected structural changes of the forest-based sector remains scarce. The uncertainties in the future outlook of the forest-based sector also imply great uncertainties in the demand for roundwood globally, and by extension, the extent of trade-offs between different ecosystem services and land uses.

Keywords Demand · Foresight · Forest-based-products markets · Forest-sector modelling · New forest-based products · Structural change

E. Hurmekoski (✉)
University of Helsinki, Helsinki, Finland
e-mail: elias.hurmekoski@helsinki.fi

L. Hetemäki
European Forest Institute, Joensuu, Finland

Faculty of Agriculture and Forestry, University of Helsinki, Helsinki, Finland

J. Jänis
University of Eastern Finland, Joensuu, Finland

© The Author(s) 2022
L. Hetemäki et al. (eds.), *Forest Bioeconomy and Climate Change*, Managing
Forest Ecosystems 42, https://doi.org/10.1007/978-3-030-99206-4_4

4.1   Background: Forest-Based Sector Outlook Studies

The outlook for the forest-based sector has great importance through the impacts that the sector has on the state of forests and the amenities that forests provide, such as forest-based products,[1] energy, employment, biodiversity, the carbon cycle and water management. Without understanding the demand for forest-based products, and the ensuing demand for roundwood, it is very difficult to assess, for example, the impacts that strategies and policies may have on the forest-based sector or on society. Nor is it possible to assess the future state of the forests.

The forest-based sector has a long history in producing *outlook studies*, extending back to the 1950s (United Nations Economic Commission for Europe/Food and Agriculture Organization [UNECE/FAO] 2021). The purposes of forest- sector outlook studies have been to examine long-term economic, social, institutional and technological trends to support policy and strategy planning, depict the range of choices available, and describe the alternative scenarios that might arise as a result of these choices (UNECE/FAO 2011). The focus has traditionally been on trends in the forest-based-products markets and the availability of wood resources, concluding that the demand for forest-based products is expected to continue to steadily increase, which results in a steady increase in the level of harvesting (e.g. Mantau et al. 2010).

In recent UNECE/FAO outlook studies, the focus has been more on 'what if' analyses, describing the potential impact of, for example, changes in the wood supply of, or demand for, forest-based products (UNECE/FAO 2011, 2021). The most-recent outlook study took the perspective of structural changes and their impacts across the global forest sector, including climate-change mitigation and adaptation (UNECE/FAO 2021). This broadening perspective is necessary in order to meet the changing information needs of policy-makers and stakeholders in the increasingly complex forest-based sector.

Indeed, there are major structural changes associated with the stagnating or declining demand for some of the traditional forest-based products, such as graphic papers and sawnwood. However, a number of innovations are also expanding the product portfolios of the forest-based industries. These changes may be the largest structural changes in a century, comparable to the uptake of wood fibres to replace rags in paper-making in the late nineteenth century. However, the methodological approaches of long-term outlook studies were adopted in an era of constant growth, and are now facing difficulties in capturing the changes taking place in the forest-products markets of the twenty-first century. Due to the lack of research and the ongoing structural changes, the outlook for forest-based-products markets remains in many ways a great unknown.

---

[1] 'Forest-based product' refers to all products made from wood raw materials and can be used interchangeably with the concepts of 'forest product', 'wood-based product', 'forest-based bio-product', etc. As the term 'wood product' may sometimes refer to solid wood industries specifically, here we use the term 'forest-based product' for consistency.

The purpose of this section is to introduce and assess some of the prominent trends and recent changes in the forest- based sector, and to examine their implications in relation to the future outlook. This section does not provide a systematic outlook, but synthesises the current knowledge and raises questions to guide future endeavours.

Box 4.1 Why Do We Need Future-Oriented Market Research?

The forest-based-products markets of the twenty-first century differ significantly from their twentieth century counterparts. The forest-based sector, as well as the operating environment, has become more fragmented and unpredictable. The customary market structures are gradually evolving, due to, for example, the diversification of product portfolios and value chains, diminishing industry boundaries, changing consumption patterns, and strengthening environmental values in business and society. This creates a need to look ahead, but at the same time, it makes this task evermore challenging.

Obviously, there is no way to directly study the future, since hypotheses regarding the future cannot be validated in the present. This is why academic future-oriented research is mainly not about predicting what is going to happen, but rather evaluating what could happen, and what would be the consequences if it did (i.e. 'what if' analysis), as well as what should happen to reach certain outcomes. Together, these guiding questions refer to *probable*, *possible* and *preferable futures*, which are considered to be the foundations of futures studies (Bell 2003).

In one of the pioneering market foresight studies, an argument and accompanying evidence were presented for the case that the rise in electronic media would significantly impact the communication-paper market (i.e. newsprint) in particular (Hetemäki 1999). The newsprint markets in North America have plummeted further and more rapidly than the study anticipated, yet at the time of publishing, these early warning signals were commonly ignored. Moreover, the study drew attention to the inability of the prevailing long-term outlook studies to capture the structural change in the newsprint markets.

By 2020, these structural changes have become evident. This poses challenges for research, as the conventional models used for long-term projections no longer sufficiently capture the market drivers, such as the factors driving substitution or the demand for new forest-based products.

In future-oriented research, it makes sense to pursue multiple approaches to obtain as comprehensive a picture as possible. One increasingly popular approach has been normative in nature—defining the means to reach set targets (e.g. backcasting). As it is not the role of a researcher to set value-laden goals, such research has to be participative. Moreover, the targets are often largely accepted, such as implementing the Paris Agreement or the UN Sustainable Development Goals. This is why future-oriented market research should increasingly be coupled with environmental impact assessment as a

(continued)

Box 4.1 (continued)

means of grasping the role and potential of the expanding forest sector in the transition to a more sustainable society.

The purpose of academic market research is to critically examine established thought patterns, present new questions, indicate knowledge gaps, unveil broader contexts, and to evoke justified views on probable, possible and preferable futures and their implications. In academic research, the methods and data need to be transparent, and the studies need to be repeatable and able to pass the peer-review process. However, the major concern with regard to the academic research on forest-based-products markets is the lack of it (Hetemäki and Hurmekoski 2016). The subject area is dominated by consulting company studies, for which there is certainly a demand, but they are not a substitute for academic research. As shown in this chapter, there are many important open questions associated with future market developments and their impacts. Thus, there is a clear need for research related to the outlook for forest-based-products markets.

4.2 Forest-Based-Products Markets in the Bioeconomy Era

4.2.1 Forest-Based-Products Markets in the Twenty-First Century

Forest-based-products *markets* refer to all industrial activities around the use of wood. Of the global growing stock of 531 billion m^3 (FAO 2018), only around 4 billion m^3, or 0.75%, is annually harvested, around half of which goes to industrial uses and half to energy (Table 4.1). The production value of the industry and energy use of wood was estimated to be approximately US$950 billion in 2018, based on FAOSTAT data. This compares, for example, to the entire global turnover in the textile industry, or the sum of the revenues of the following companies in 2018: Apple, Amazon, General Motors, Microsoft, Bank of America, IBM and General Electric. Importantly, these figures only refer to the core industrial activities and do not include various downstream industries and related services. Clearly, the forest sector plays a significant role in the global economy and employment, besides heavily influencing the state of the world's forests.

Wood is used for various purposes, such as for buildings, furniture, packaging, communication, decoration, clothing, hygiene, vehicles, paints, glues, detergents, fuel, heat, medicine, feed and food. Forest-based industries are typically separated into the solid-wood industries, comprising sawnwood, wood-based panels, furniture and engineered wood products (EWPs), and the chemical forest industries, comprising pulp, paper and paperboard. There is little in common between the solid and chemical forest industries, save for the raw material supply.

Table 4.1 Global forest-products production in 2018

	Production quantity *(million tons)*	Production value *(billion US$)*
Industrial roundwood	1014*	–
Wood fuel	972*	146
Paper and paperboard	409	374
Wood pulp	188	137
Sawnwood	246*	135
Wood-based panels	204*	159
Total	–	950

* Converted from m³
Source: FAOSTAT

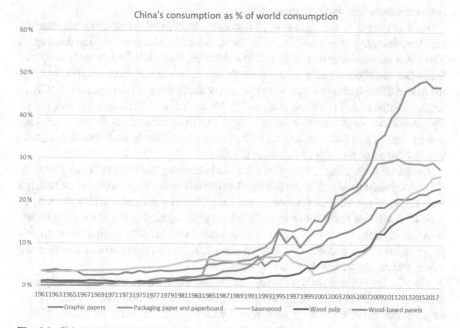

Fig. 4.1 China's share of the global consumption of major forest products. (Data: FAOSTAT)

There have been many visible changes in the global forest-products markets in the twenty-first century. For example, the global competitive advantages have experienced a clear shift, with a remarkable share of forest-industry investments going to fast-growing markets in Asia and low-cost-production regions, such as South America. The increase in demand for forest products in the 2000s originated almost entirely in Asia (FAOSTAT), with China's share of the global consumption having grown to more than 20% of all major forest products by 2018 (Fig. 4.1).

A more profound, yet less tangible, change is the *structural change* in the demand patterns of the forest-based industries. In the twentieth century, the global demand for forest products was steadily increasing, driven by increasing incomes, populations and urbanisation. However, in the twenty-first century, many of the forest-based products no longer seem to follow the pattern of stable and predictable growth of the last century. This is a consequence of various structural changes in demand, driven by the substitution of forest-based products for, or by, competing products. For example, global graphic- papers production ($\approx$ consumption) declined between 2007 and 2018 by almost a quarter (24%), according to FAOSTAT data, due to its substitution by electronic media. On the other hand, in the same period, the production of dissolving pulp has grown by 2.5 times, driven by textile industry needs.

Important drivers of structural change also include the Sustainable Development Goals (UN 2015a) and the Paris Climate Agreement (UN 2015b). These set internationally agreed goals that encourage sustainable production and consumption. As forests constitute the most important, non-food, renewable land resource, increasing interest in utilising forests as a substitute for fossil-based and other non-renewable feedstock materials can be expected.

As a response to the maturing or declining traditional forest-products markets and the emerging opportunities, new forest-based products are being developed. Thus, it is conceivable that, within a few decades, there will be a larger number of forest-based-products categories, although none of these will dominate the sector to the extent that paper and solid- wood products did in the last century, particularly in terms of value added (Jonsson et al. 2017). Moreover, with the new products, industry boundaries may become increasingly indistinguishable, with the chemical, energy, textile and forest industries using the same feedstocks and developing products for the same markets (Jonsson et al. 2017).

Based on these trends, a keyword for characterising the market development of the forest-based products in the twenty-first century is *diversification*. Above all, the term refers to the widening scope of the forest-based-products markets in terms of product portfolios and value propositions. One can argue that the sawnwood industries are diversifying towards wood- based panels and EWPs, whilst the pulp and paper industries are diversifying from communication papers towards packaging paper grades and various biorefinery products.

Clearly, the outlook for the forest-based sector depends on whether we only consider the traditional large-volume products, such as sawnwood and graphic papers, or also the development of *new forest-based products*, such as textile fibres. Capturing the influence of the latter can be tricky, as sectoral statistics are lagging behind the restructuring of this industry. In particular, it is increasingly challenging to measure the employment, turnover and value added based on wood raw materials in the chemical, construction, textile and energy industries.

4.2.2 Characterising the Structural Change in Demand

Industrial evolution is a continuous process that serves to maintain the vitality of the market economy, as already noted in the 1940s by Joseph Schumpeter, who coined the term '*creative destruction*'. Here, we briefly introduce a few analytical concepts so as to characterise the structural changes occurring in the forest sector in the twenty-first century—evidence of ongoing creative destruction.

4.2.2.1   Demand Elasticity and Substitution

By the term '*demand*', we refer to the amount or value of a good consumed. A useful empirical approximation of demand is 'apparent consumption', defined as production + imports – exports. The terms 'demand' and 'consumption' are therefore regarded as synonyms.

As most forest-based products are intermediate goods, models quantify forest-based-products consumption as *derived demand*. Essentially, this means that the demand for forest-based products is a function of the same factors that affect the demand for the final uses of the products (Klemperer 2003). In empirical research, the demand determinants for wood- based products are typically reduced to price and income.

One way to demonstrate the existence of a structural change is to observe the relationship between available income and the demand for forest-based products. This leads us to the concept of demand *elasticity*. While elasticities can be attributed to any demand determinant, they are typically associated with price and income. The demand for a *normal good* increases when income increases and decreases when income decreases, whereas, for an *inferior good*, an increase in income results in a reduction in this consumption and vice versa (Varian 2010). Forest-based products are generally regarded as normal goods (Kangas and Baudin 2003), except for newsprint and printing and writing papers (Hetemäki 2005).

Indeed, the income and price elasticities remained remarkably stable throughout the twentieth century, when the markets enjoyed a period of relatively stable growth, and there were no major technical innovations making competing goods more desirable. Consequently, income and price have been able to explain and predict the level and rate of demand remarkably well at the global level. However, the power of these two predictors diminish, the more disaggregated markets and more recent data are analysed.

In the twenty-first century, the demand for forest-based products has no longer developed in line with the gross domestic product (GDP) for some of the most significant forest products. This suggests that income cannot be the only demand shifter. For example, the global production of sawnwood and wood-based panels has exhibited markedly different patterns in 2000–2018 compared to 1980–1999, relative to per-capita GDP growth (Fig. 4.2). In the EU, one can observe an apparent decoupling of demand from the GDP, or a structural break or discontinuity in the GDP elasticity, for many traditional forest-based-products markets (Fig. 4.3). The underlying causes have been studied only in the context of graphic papers, which are being substituted by electronic media (Hetemäki 1999; Hetemäki and Obersteiner 2001), and bioenergy, which has been substituting for fossil energy in the EU due to climate and energy policies (Moiseyev et al. 2013). Despite some of this apparent turbulence possibly being caused by the historically long economic downturn, therefore making it transitory, it may equally become more exaggerated in the future due to the introduction of new forest- based products and new end uses, as well as the impacts of the COVID-19 pandemic.

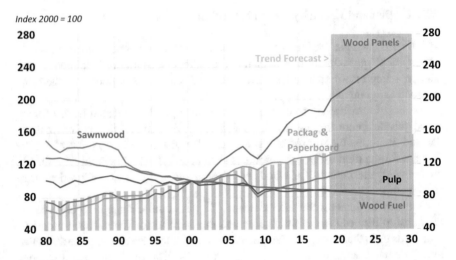

Fig. 4.2 Per-capita global consumption of forest products versus GDP in 1980–2018 and the trend forecasts (2010–2018 trend) to 2030. (Data: FAOSTAT and World Bank)

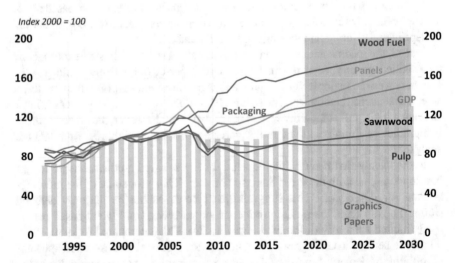

Fig. 4.3 Consumption of wood-based products and GDP in Europe (excluding Russia) in 1992–2018 and the trend forecasts (2010–2018 trend) to 2030. (Data: FAOSTAT and World Bank)

The lack of research literature on *substitution*, in the context of forest-products markets, is striking, given the structural changes that took place in the graphic papers markets in many Organisation for Economic Co-operation and Development (OECD) countries some decades ago (recent exceptions being described in Latta et al. 2016; Rougieux and Damette 2018). Moreover, the commonly stated goal of 'shifting towards bioeconomy' would implicitly require the large-scale substitution of feedstock materials, and yet the conventional demand equations cannot fully capture this substitution. The implications of this knowledge gap are further underscored in Chap. 7.

4.2.2.2   Product Life-Cycles

The diversification of the sector is closely related to the *product life-cycle*, comprising four to five stages— introduction, growth, maturity, decline and, in some cases, renewal (e.g. Routley et al. 2013). During the period of introduction and growth, the goods become increasingly competitive through decreasing production costs from learning- by-doing (Arrow 1962; Rosenberg 1982). At the maturity stage, productivity improvements are increasingly difficult to gain, and in the decline stage, the product starts to lose the markets to emerging products or technologies (Anderson and Tushman 1990). In some cases, the growth phase may be renewed after a period of stagnation or decline, as a result of changes in demand determinants or improvements in the established technology—this has been the case for dissolving pulp, for example.

The demand for most woodworking and pulp and paper industry products has become inelastic; that is, the market growth rate has fallen below the GDP growth rate (Rougieux and Damette 2018). Moreover, the prices of the end products have been trending downwards, while the production costs have been increasing, with the price differentials between suppliers being marginal, the switching costs being low, and the negotiating power lying with the customers (Uronen 2010; Hetemäki et al. 2013). These point to the conclusion that many of the forest-products markets have become commoditised. At the same time, in the big picture, fossil-based energy is likely to hit the maturity stage in the coming decades, due to environmental values and regulations, which will lead to substitution by alternative energy and material feedstocks, including forest biomass.

Indeed, one can name wood-based products for all phases of the typical product life-cycle, as demonstrated in Fig. 4.4. For example, cross-laminated timber (CLT) is clearly in the growth phase, as demonstrated by the double-digit growth rates, irrespective of periods of negative or stagnating GDP growth rate (Hetemäki and Hurmekoski 2016). In the next section, we focus particularly on the markets of emerging wood-based products.

4.2.3   Emerging Markets

4.2.3.1   Defining New Forest-Based Products

There is no clear or established definition for new forest-based products (Cai et al. 2013; Näyhä et al. 2014; Hetemäki and Hurmekoski 2016). The concept can refer to products in the introduction or growth phases of the product life-cycle, but also to products with renewed growth in demand, such as dissolving pulp. Thus, a new forest-based product does not necessarily have to be a novel product, or based on a novel technology—it can also be an old product with a new market environment, such as the constrained supply of competing feedstock materials, or with enough incremental improvements to drive growth, such as a lighter weight (Hurmekoski

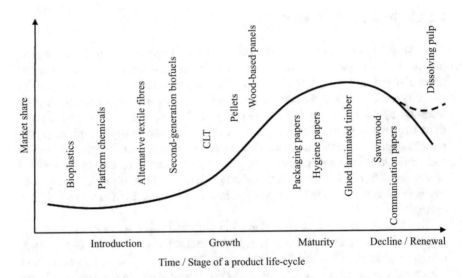

Fig. 4.4 Approximate market location of selected wood-based products based on product life-cycle in 2020

et al. 2018a). Thus, one could describe new wood-based products as products for which the demand is determined mainly by drivers other than economic activity (GDP), bar short-term economic cycles. Such a classification is naturally prone to interpretation in terms of where to draw the line. For example, compared to CLT and dissolving pulp, the growth of the fibre-based packaging sector, as a whole, is not necessarily fast enough to qualify as a new forest-based product, even if some of the emerging packaging applications are novel, such as cups containing no fossil plastics. These definition attempts may at least demonstrate the diversity of the forest-based- products markets in the twenty-first century.

Literature on the diversification of the sector and new forest-based products remains relatively scarce, particularly from the perspective of market potential (e.g. Guerrero and Hansen 2018). Based on the literature that is available, however, some of the most important emerging markets appear to be construction, textiles, chemicals, advanced biofuels, and plastics and packaging (Bio-based Industries Consortium 2013; Graichen et al. 2016; Antikainen et al. 2017; Kruus and Hakala 2017; Schipfer et al. 2017). In the following, we briefly review these markets, as summarised in Table 4.2.

4.2.3.2 Construction Markets

The outlook for wood construction is regarded as almost unanimously positive, which is partly due to the enormous and still-growing size of the market, and partly due to claims of the superior environmental performance of wood products (e.g. European Commission 2018). Wood construction, particularly on a small scale,

Table 4.2 Markets for new forest-based products

	Construction	Textiles	Chemicals	Fuels	Plastics and packaging
Market size in 2030 (*in 2015*)	28,000 Mt. (*21,500 Mt*); 3.16 billion m² (*2.24 billion m²*)	130 Mt. (*90 Mt*)	600 Mt. (*330 Mt*)	2300 Mt. (*2100 Mt*)	130 Mt. (*72 Mt*)
Technologies / products	- EWPs (CLT, laminated- veneer lumber) - industrially prefabricated construction elements (including modular elements) - concrete admixtures (lignin)	- new solvents for dissolving pulp: e.g. IONCELL-F - new fibre-spinning technologies: e.g. Spinnova	- drop-in substitutes for petrochemicals: Ethylene - smart drop-in substitutes for petrochemicals: Succinic acid, butanediol - dedicated bio-based chemicals: Lactic acid, furfural	- renewable diesel: Based on distilling tall oil - ethanol: Based on fermenting sugars (hemicelluloses and celluloses)	- wood–plastic composites (WPCs): Extrusion and injection moulding - pulp-based, paper-resembling films for flexible packaging - other plastic-resembling wood or wood-fibre-based materials for rigid packaging
Target markets and substitutions	- residential and non-residential buildings - substituting for concrete, steel and established wood-construction technologies in the load-bearing frames of buildings	- garments - substituting cotton, polyester and viscose	- main downstream markets, including plastics, food and feed ingredients, and pharmaceutical industries - substituting first-generation (starch-based) biochemicals and petrochemicals	- energy carrier for transport, particularly long-haul truck transport, maritime transport and jet fuel - substituting first-generation biofuels and fossil fuels	- rigid and flexible plastic substitutes: Food, healthcare and cosmetics packaging, carrier bags - WPCs: Decking (mostly substituting for tropical wood), car interiors (mostly substituting for plastics)

(continued)

Table 4.2 (continued)

	Construction	Textiles	Chemicals	Fuels	Plastics and packaging
Main drivers	- efficiency gains in industrial prefabrication - favourable policies in certain regions	- 'cellulose gap'—Constrained farming area for cotton due to land competition with food production, coupled with rapid growth in demand for textiles - large freshwater consumption in cotton irrigation in arid areas	- major firms targeting renewable feedstocks - co-production with biofuels	- climate and energy policies - crude oil and CO_2 price in the long run	- growth in population, GDP, e-commerce and take-away products - rising polymer prices - policies to restrict the use of plastics
Main barriers	- risk perceptions of key decision-makers (officers of main contractor and developer firms) - fragmented and path-dependent industry structure	- some of the technical attributes (product properties) of man-made cellulosic fibres (MMCFs)	- REACH[1] and other regulations - extensive validation required for dedicated compounds - investment costs - path dependency of petrochemical industries	- feedstock availability - for many processes, conversion efficiency - investment and running costs	- uncertain legislative environment
Competing innovations	- low-emissions cement - 3D printing of recycled concrete and similar	- cotton recycling technology - other bio-based fibres, e.g. based on spiders' webs - functional textiles, e.g. antibacterial, anti-odour or electrical properties	- CO_2 as a feedstock for chemicals - first- and third-generation chemicals	- first- and third-generation fuels - electric engines - hydrogen engines	- first- and third-generation bioplastics - recycled or biodegradable plastics - natural fibre composites

Desirable product characteristics	- no need for major changes in construction practices - no technical or economic hazards	- technical properties: Avoiding wrinkles and electricity, good moisture absorption, etc. - environmental properties: Avoiding hazardous chemicals, less pollution, increased recycling, etc.	- low cost - non-hazardous and non-toxic	- drop-in fuel: Existing distribution infrastructure and existing car fleet without a need for major modification	WPC: - natural feel - easy maintenance packaging: - biodegradability or recyclability - lightness - product safety
Comparative advantages	- lightness of the material, allowing efficient industrial prefabrication and the resulting productivity benefits - renewable material	- feedstock availability (compared to virgin cotton) - ability to convert existing pulp mills to dissolving pulp - environmental footprint	- interest towards bio-based alternatives - in smart drop-in and dedicated chemicals, reduced costs and/or environmental footprint	- policy pull - does not directly compete with food production - can be biodegradable, free of aromatics and Sulphur, and non-toxic	- reduced costs compared to pure plastics - combination of biodegradability and thermoplasticity
Position of forest-based firms in the value chain	- admixture supplier - subcontractor (product or element supplier) - main contractor or developer (managing whole value chain)	- raw material supplier (dissolving pulp) - textile fibre producer (MMCFs) - yarn producer	- primary and secondary platform chemicals	- end-product producer	- packaging: Converter of shopping bags and solid packages - WPC: Converter of intermediate/end products

^aNote: REACH—Registration, evaluation, authorisation and restriction of chemicals—is an EU regulation covering the production and use of chemicals
Adapted from Hurmekoski et al. (2018a)

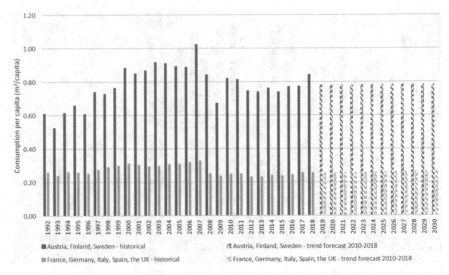

Fig. 4.5 Per-capita consumption of sawnwood and wood-based panels in two selected country groups in the EU in 1992–2018. (Data: FAOSTAT)

such as single-family homes, can hardly be regarded as a new market, per se, but new opportunities are emerging in the large-scale construction markets due to recent innovations, as well as growing interest among decision-makers and industries (e.g. FAO 2016). The market can also be approached from an entirely different perspective, such as using lignin to partly replace cement in concrete manufacturing (de Vet et al. 2018).

The average market share of wood in small-scale construction in Europe has remained below 10%, but it varies from above 80% in the Nordic countries to near zero in many Southern European countries (Alderman 2013). Figure 4.5 shows that, during the last couple of decades, the per-capita consumption of construction-related forest-based products in the large European economies has only been around one-third of the equivalent consumption in sparsely populated and densely forested countries. Furthermore, there is no convergence in the per-capita consumption between the high- and low-consumption regions, with a slight exception being modest growth in the UK. Figure 4.5 also indicates that the per- capita consumption of wood-based construction products in Austria, Finland and Sweden nearly doubled from 1993 to 2007, after which the markets were severely affected by the global economic downturn, and did not manage to reach the peak of 2007 in the decade that followed. Looking at the trend forecast based on the trend of the 2010s, no significant deviation would be expected.

Besides the long-lasting impact of the housing-market meltdown, the lack of significant progress in the countries with a smaller wood-construction market share arises from the various path dependencies of the construction sector (e.g. Mahapatra and Gustavsson 2008). Construction markets are very much influenced by tradition, culture and the availability of local resources. In this highly established market, major drivers include cost competitiveness and being able to guarantee a low-risk investment. That is, particularly in large-scale construction value chains, the actors are generally unwilling to accept new practices that could potentially cause extra work and associated costs in the short run (Arora et al. 2014). There is, however, significant variance in the market potential between market segments and regions, even from one city to the next.

The demand drivers are slightly different between the small-scale and large-scale housing markets. Some of the technological- and business-model-related innovations hold promise for changing the market prognosis in the large-scale construction markets. In particular, the expansion of *EWPs*, together with *industrial prefabrication*, has allowed wood to increasingly compete with steel and concrete in large-scale construction (Bühlmann and Schuler 2013; Hildebrandt et al. 2017), such as in multi-family dwellings, office and industrial buildings, sports complexes, additional-storey construction, and in infrastructure, such as bridges. Industrial prefabrication refers to the off-site manufacturing of elements and components, which allows the combination of several work phases in a single off-site location, potentially resulting in productivity and quality gains, for example (Malmgren 2014). However, even if wood-based industrial prefabrication could address many of the pressures faced by the construction sector, including productivity, quality, safety and environmental impact, the risks, as perceived by the construction project managers, might outweigh these in the short term (Hurmekoski et al. 2018b). These hindrances would need to be addressed, for example, by taking responsibility for a larger share of the construction value chain, if firms seek rapid market growth (Hurmekoski et al. 2018b). More-gradual change can happen through standardisation and winning trust through repeated positive experiences along the value chain (Hurmekoski et al. 2018b).

The market is also influenced by policy. In Finland, wood construction has been promoted by public targets, technology platforms and campaigns for several decades. While the small-scale construction markets have already been saturated by wood, the uptake of wood-frame, multi-storey construction remains modest, with only a few percent market share. In the 2020s, the possible uptake of environmentally stricter national regulations, driven by, for example, the national implementation of the voluntary EU framework for measuring the emissions of the construction sector or supportive measures favouring wood in public procurement in the building sector, may favour wood (Toppinen et al. 2018), besides spurring competition.

Changing consumer preferences may also influence the market uptake of modern wood-construction practices. For example, wooden surfaces may have beneficial impacts on human health through improved air quality and a stress- relieving atmosphere (Muilu-Mäkelä et al. 2014), which has been found to be attractive to certain types of consumers (Lähtinen et al. 2019). However, particularly in multi-storey buildings, consumers tend to emphasise the size, location and price of the apartments rather than the material of the structural frames. Here, consumer segmentation and business model innovations may be required.

4.2.3.3 Textile Markets

The global textile demand has been projected to grow from 90 Mt. in 2015 to more than 250 Mt. in 2050 (Alkhagen et al. 2015). In the absence of more-efficient recycling, polyester fibres are foreseen as having the strongest growth, followed by a more stable increase in cellulosic fibres, while cotton production is expected to remain at the current level (Antikainen et al. 2017), due to the increasing competition for land between cotton and food production (Hammerle 2011). Additionally, the large demand for fresh water for cotton farming in arid areas lends a competitive advantage to alternative cellulose supply sources (Shen et al. 2010), such as wood-based MMCFs.

The MMCF market is still dominated by viscose, with a 79% share in 2018 (Textile Exchange 2019)—a product that was introduced already in the late nineteenth century. Currently, new MMCF processes based on alternative solvents, such as IONCELL-F and Arbron, are being developed that aim to overcome the weaknesses of contemporary viscose (Kruus and Hakala 2017). Of these more advanced MMCFs, dissolving-pulp-based lyocell already had a 4% market share of the MMCF markets in 2018, and it is expected to grow faster than viscose (Textile Exchange 2019). If the development of new cellulosic fibres is successful, the general growth rate of MMCFs can be higher than so far perceived, and could expand into currently unattainable markets, such as sports textiles (Alkhagen et al. 2015).

The main intermediate product in manufacturing MMCF is dissolving pulp. Following a decline lasting the four decades since 1960, the global production of dissolving pulp has grown from 2.8 Mt. in 2000 to 8.4 Mt. in 2018, reaching 4.5% of the overall wood-pulp production volume (FAOSTAT). With a growing global textile demand, an increasing number of kraft pulp mills could be converted to produce dissolving pulp. Dissolving pulp is currently exported in large quantities to Asia, where most of the global textile production takes place. In principle, the value added of wood-based industries from the textile market could be multiplied by moving downstream in the value chain to garment manufacturing (Hurmekoski et al. 2018a).

According to Antikainen et al. (2017), the textile supply chains are typically long and complex, while the use time of textiles is relatively short due to low pricing and rapid fashion cycles. The industry is further characterised by a high share of labour costs (Antikainen et al. 2017). As a consequence, textile manufacturing has been off-shored from much of the global West to regions with lower wages—notably, the Far East countries. Even considering the possibility of highly automated textile production, Antikainen et al. (2017) did not foresee any substantial reshoring of garment manufacturing to the Western economies. Instead, the technology that is being developed could be licenced to areas where textiles are produced, and recycled cotton fibres could also be used as feedstock, along with wood pulp, for example (Kruus and Hakala 2017).

4.2.3.4   Biorefining—Biochemical and Biofuel Markets

There is a clear need to find alternative feedstock materials for fossil coal, oil and gas. Technically, almost all industrial materials made from fossil resources could be substituted by their bio-based counterparts (de Jong et al. 2012). However, so far, the most common uses of wood have been to exploit the solid wood or fibre structure, as some of the molecular structures have been too complex to be replicated by engineers, and there are cheaper sources for the development of chemicals. This setting could be slowly changing, due to constantly developing technologies, as well as changes in the *operating environment.*

Biorefinery feedstocks can be divided into three generations—the higher the generation, the less competition the feedstock poses for land use and food production, but, at the same time, the higher the techno-economic barriers for the market uptake (Sirajunnisa and Surendhiran 2016). First-generation biorefineries are mostly based on food crops or plants that reduce the land available for food production, which drives up the food price, particularly in developing countries (Naik et al. 2010). Typically, the first-generation biorefineries are also technologically less efficient and have higher carbon footprints than second-generation biorefineries (Naik et al. 2010). Second-generation biorefineries are based on lignocellulosic biomass, such as wood and agricultural residues or waste streams, and do not directly compete with food production. A third-generation biorefinery is yet to be established, but it would be based on various algae that do not require any land surface for cultivation.

So far, most market prognoses for biorefining concern the first-generation feedstocks (e.g. Aeschelmann and Carus 2015). Accordingly, the literature on lignocellulosic biorefineries tends to focus on the technical challenges related to its pretreatment phase, such as the insufficient separation of cellulose and lignin, the formation of byproducts that inhibit downstream fermentation, the high use of chemicals and/or energy, the cost of enzymes, and the high capital costs for pretreatment facilities (Taylor et al. 2015). Given that the technical-readiness level of most wood-based chemicals remains fairly low (Hurmekoski et al. 2018a), the market outlook, even up to 2030, remains uncertain, despite the obvious potential.

Advanced Biofuels

In terms of sheer volume, the substitution potential is several magnitudes higher for bioenergy than for biomaterials (Schipfer et al. 2017). According to Plastics Europe (2016), 42% of all oil and gas in Europe is used for electricity and heating, while 45% is used for transportation, 8% for chemistry (plastics 4–6%) and 5% for other uses.

The demand for biofuels has largely been created by national or regional climate and energy policies, such as the Renewable Energy Directive of the EU, which necessitated a 10% blend of biofuels in road traffic by 2020 (EU 2009). However, the first-generation biofuels have faced criticism due to the uncertainties associated with their ability to reduce emissions, their low conversion efficiency, and the low energy return on energy invested in some of the processes (de Jong et al. 2012). This has created interest in advanced biofuels based on second-generation feedstocks.

The investment requirements can be up to three times higher for lignocellulosic biofuels than for cornstarch- or sugarcane-based ethanol (Nguyen et al. 2017), due to there being more steps in the production process. Also, the C5 and C6 sugars produced in lignocellulosic biorefineries for fermentation are much more expensive than sugar from sugarbeet or sugarcane (Carus et al. 2016). The cost disadvantage of sugars derived from lignocellulosic feedstocks would need to be balanced by the utilisation of lignin, which seems to be feasible only in the very long term, in terms of its full potential (Carus et al. 2016). For this reason, the production of biofuels typically requires the complementary production of biochemicals (or selling the residues for such use) to make the business profitable. Indeed, the value of the chemical industry is comparable to that of the fuel industry, despite it requiring only a fraction of the biomass (FitzPatrick et al. 2010).

Chemicals

In Finland, more than a third of the chemical industry firms use bio-based feedstocks, and the number is expected to rapidly increase (Ministry of Employment and the Economy 2014). The chemical industry is therefore seen as playing a key role in the diversification of the forest-based sector.

Like biorefineries, chemicals can be classified in many ways. In terms of markets, an important distinction is for bulk chemicals (or basic chemicals), characterised by a high volume, low price and highly diverse end uses, and fine chemicals, characterised by a low volume, high price and few applications. The mixtures of the chemicals in these categories constitute specialty chemicals (or performance chemicals), including adhesives, agrichemicals, detergents, cosmetic additives, construction chemicals, elastomers, emulsifiers, flavourings, food additives, fragrances, industrial gases, lubricants, pigments, polymers and surfactants. A probable role for the forest-based industries would be to supply basic or fine chemicals to be sold to the chemical industries for a plethora of end uses (Hurmekoski et al. 2018a).

In terms of volume, the chemical sector is dominated by a small number of key bulk chemicals, such as ethylene. Producing bio-sourced, drop-in basic chemicals with an identical compound structure to their fossil-based counterparts appears to be a promising approach for the biochemical markets, due to a greater likelihood of acceptance by established chemical producers (FitzPatrick et al. 2010). That is, a functional replacement with a different molecular structure would require significant property testing to allow displacement of the chemical currently being used (Biddy et al. 2016). However, while drop-in, bio-based chemicals are seen to have an easier access to the markets compared to other types of chemicals (de Jong et al. 2012), they are not expected to be competitive by 2030 due to their low technology readiness, longer conversion pathways and comparably high running and investment costs (e.g. Kruus and Hakala 2017). As a consequence, Bazzanella and

Ausfelder (2017) recommend exploiting the more efficient synthesis of target products that maintain the functional units of the feedstock molecules, such as polylactic acid.

Let us consider the case of ethylene to highlight the complexity of the interplay of market drivers and barriers in the chemical market, and the resulting difficulty of assessing the market potential. According to Dornburg et al. (2008), ethylene (used mostly for polyethylene plastic) is the largest of the currently produced petrochemicals by volume. While bio-based ethylene has a technical readiness level of 8–9 out of 9, it is competing with natural gas and, particularly, shale oil and gas, which have higher relative yields of ethylene compared to conventional oil and gas sources (Biddy et al. 2016). Yet ethylene production could fit into the overall product portfolio of wood-based biorefineries, if certain parts of the feedstock would otherwise have no use. If there is a price premium for bio-sourced ethylene, even a minute share of the global market could have a large impact on the profitability of a single biorefinery.

Unlike for biofuels, the demand shifters of biochemicals are not only related to policy. There has been a shift from a technology push led by major chemical companies to a market pull created by leading consumer brands, such as P&G, IKEA, LEGO and the Coca Cola Company, which all have set specific targets for replacing fossil-based chemicals (polymers) with more sustainable alternatives (Aeschelmann and Carus 2015; Biddy et al. 2016). Naturally, policies could also create incentives or influence the relative costs of different feedstocks through, for example, CO_2 pricing to level off the differences between the operating and investment costs. Due to the myriad end uses of chemicals, the market drivers and consumer preferences may also vary significantly from one use to another.

According to Hurmekoski et al. (2018a), wood-based chemicals may primarily compete with first-generation biochemicals and chemicals produced from other second-generation feedstocks rather than petrochemicals, which would lower the expected production volume considerably. However, it is too early to state any such prognoses with certainty, and there seem to be ripe opportunities already available, based on the €550 million standalone wood-based chemical mill investment by UPM Kymmene in Germany in 2020, which will be producing dedicated biochemicals for the production of items such as textiles, bottles, medicines, cosmetics and detergents. Beyond 2030, the competition may change again with the introduction of, for example, CO_2 as a feedstock for the development of platform chemicals (Alper and Orhan 2017).

Besides cellulose and hemicellulose, wood also contains lignin and a number of heterogeneous extractives, derived, for example, from birch bark. There are highly varied opportunities for such niche markets, resulting in a large number of speculative uses in the long term (Box 4.2). Due to the numerous opportunities and the early stage of their life-cycle, we can only argue that they may eventually make a big difference in terms of value added, albeit the volume potential remains fairly low due to the restricted availability of byproducts, except for lignin.

Box 4.2 Opportunities Related to Byproducts and Niche Markets
The biofuels and biochemicals reviewed in this section are mostly based on fermenting or catalysing *C5 and C6 sugars* from cellulose and hemicellulose. Besides cellulose and hemicellulose, wood also contains lignin and a number of heterogeneous extractives (Fig. Box 4.1).

Around 70 Mt. of *lignin* are produced annually in the world, with 95% of the production being incinerated. Other uses include dispersants (e.g. in the construction industry), emulsifiers (asphalt emulsions), foams (plastics/polymers), aromatics (vanillin) and stiffness enhancers (corrugated board) (Bruijnincx et al. 2016), with most utilised, without chemical modification, as fillers or additives (Aro and Fatehi 2017). However, lignin is considered to be a main aromatic renewable resource for the development of chemicals and polymers in the long term (Laurichesse and Avérous 2014). The low rate of lignin exploitation to produce chemicals is mostly due to its complex, largely undefined structure and its versatility depending on the origin, as well as its tedious separation and fragmentation processes (Laurichesse and Avérous 2014). That is, while the limitations on the use of hemicellulose relate to markets rather than technology (Stern et al. 2015), the opposite holds true for lignin (Bruijnincx et al. 2016). Platform or fine chemicals based on the thermochemical conversion of lignin show a low technological-readiness level (Kruus and Hakala 2017). However, the possibilities are highly varied, resulting in a countless number of speculative uses in the long term.

Wood also contains several small and highly heterogeneous extractives, such as *terpenes*, *fatty and resin acids*, *sterols*, *phenolic compounds* and *hydrocarbons*, which hold considerable potential as renewable resource s and feedstocks for future biorefineries (Routa et al. 2017). For example, tree bark

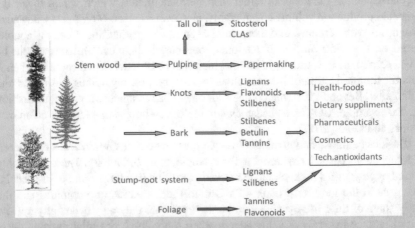

Fig. Box 4.1 Currently utilised and potential routes from wood extractives to valuable biochemicals. *Green arrows* commercialised routes, *red arrows* non-commercialised routes, *CLAs* conjugated linoleic acids. (After Routa et al. 2017)

(continued)

Box 4.2 (continued)

contains bioactive components, such as tannins—phenolic compounds that could be used in several technochemical applications and products, such as glues, wood preservatives, foams, functional coatings and adhesives. Tannins have widespread applications in many other industrial sectors, including the food, beverage, clothing and pharmaceutical industries (Shirmohammadli et al. 2018). Bark also contains non-cellulosic sugars, which can be utilised further, but also make direct tannin extraction difficult (Kemppainen et al. 2014).

A very interesting group of compounds are *the birch wood extractives*, which possess considerable potential utility. Natural birch bark extractives, such as triterpenoids (e.g. betulin and related derivatives), suberinic fatty acids and phenolic compounds, find potential use in pharmaceutical, techno-chemical and food/feed applications. Birch bark triterpenoids have shown a marked potential as precursors for HIV and cancer therapy, as well (Krasutsky 2006). For the recovery of wood extractives, thermochemical techniques, such as pyrolysis, hot-water extraction and hydrothermal liquefaction, seem to be the most promising technologies. Nevertheless, large-scale operations for the recovery and further refinement of wood extractives are still scarce.

The utilisation of *tall oil* is mainly focused on the production of renewable diesel. However, crude tall oil refining also produces side-products, such as tall oil rosin and tall oil pitch, which find use in many different applications, especially in the technochemical field (Routa et al. 2017).

Cellulose can also be broken down to the *nanoscale*, which alters the properties of the fibres, giving them superior strength, liquid-crystal behaviour, transparency, low thermal expansion, the capacity to absorb water, and piezo-electric and electrical behaviours (Cai et al. 2013). According to Cowie et al. (2014), the largest uses for nanocellulose are projected to be in packaging (2 Mt), paper (1.5 Mt) and plastic film (0.7 Mt) applications. Globally, the use of nanocellulose as a cement additive has a potential market size of over four million t. Other applications include functional paper, coatings, packaging, pharmaceuticals, cultivation media, biosensors, various membranes, catalysts, polymer composites, textiles and electronics (Thomas et al. 2018). So far, larger-scale applications are limited to increasing the strength and reducing the weight of conventional carton packaging, for example.

4.2.3.5 Plastics and Packaging Markets

Similar to the construction market, there is a long tradition of using wood in the packaging market s. Generally, the packaging markets are driven by global population and GDP growth, as well as increasing e-commerce and the demand for take-away products. However, wood-based packaging solutions may have increasing potential due to an increasing resistance against plastics, originating particularly from marine and microplastic pollution (World Economic Forum 2016). For example, in the EU, certain short-lived plastic products have been banned, and the use of plastic bags is being disincentivised, creating market pull for alternative materials. The packaging markets also represent one of the most important uses of bio-based chemicals (Hämäläinen et al. 2011; Näyhä and Pesonen 2012). The global production of plastics has increased 20-fold over the past 50 years, from 15 Mt. in 1964 to 311 Mt. in 2014 (World Economic Forum 2016). Over the next 20 years, the volume is expected to double, and by 2050 to quadruple (1.124 Bt) (World Economic Forum 2016). Of the total global plastic market in 2015 (322 Mt), 40% ended up in packaging, while up to 70% of bioplastics are used for packaging (Plastics Europe 2016). Plastics, and paper and paperboard each account for around 35% of the total value of the packaging markets (Neil-Boss and Brooks 2013). The bioplastics market is expected to gain a 5% share of the entire plastic-packaging market within 20 years (Byun and Kim 2014), but the share of wood-based polymers from the entire bio-based polymer market remains modest.

As noted by de Jong et al. (2012), a plastic with a technical function and complex supply chain could take between two and four decades to achieve production scales over 100,000 t. According to Aeschelmann and Carus (2015), novel, 100% bio-based, indirect-substitute polymers are not expected to grow as fast as the drop-in polymers until 2030. Moreover, the advantage of bioplastics in a circular economy is not clear, as they are not necessarily biodegradable, and rapid biodegradability is not necessarily beneficial for the environment either, if the material cannot be recycled (Soroudi and Jakubowicz 2013).

This leads us to argue that plastics, as such, are not necessarily a key business opportunity for the forest-based industries (Hurmekoski et al. 2018a). Combined with the technical and economic issues raised for the biochemical market, and the likely role of forest industries as a platform chemical provider, indirect-substitute products for the plastics market could have more potential by 2030. These indirect substitutes could be plastic-mimicking products that use existing industrial infrastructure, such as WPCs (Carus et al. 2015), paper-resembling films for flexible packaging (Kruus and Hakala 2017) and other plastic-resembling wood or fibre-mix materials for rigid packaging (e.g. Nägele et al. 2002). The demand for such indirect plastic substitutes could be promoted by policy, such as the EU directive that bans certain single- use-plastic products. Naturally, the demand for traditional wood-based packaging or 'second-generation' fibre-based packaging (free from fossil-based polymer coatings and adhesives) could increase their market share, vis-à-vis plastics, glass and aluminium, as long as they are compatible with the disposal and recycling behaviours in a circular production– consumption system, satisfy heterogeneous consumer needs, and support sustainable lifestyles by extending material life- cycles (Korhonen et al. 2020).

4.2.3.6   Impacts of New Forest-Based Products on the Forest-Based Sector

Despite the wide array of possibilities associated with new forest-based products, the core forest industry products— sawnwood and pulp and paper—are likely to still retain a significant role in 2030. This is due to the continuous demand for them, as well as the long process involved with introducing a new product to the markets and gaining large market volumes, and the long investment cycles of the forest-based industries. Another reason is that the research and development (R&D) and investment in new products is funded, to a significant degree, by turnover from the traditional businesses, albeit wood-based innovations, such as textiles, may arise from outside the traditional forest sector.

As local conditions and cultures vary, so do the shapes of the forest-based bio-economy business models, as exemplified by the biorefineries of Borregaard in Sarpsborg, Norway and the Metsä Group in Äänekoski, Finland (Hetemäki and Hurmekoski 2020). Borregaard focuses on low-volume global niche markets rather than commodity products, and makes relatively large R&D investments to serve, for example, the agriculture, construction, pharmaceutical, cosmetics, food and electronics markets. In contrast, the bioproduct mill in Äänekoski—the largest forest-industry investment in the Nordic countries—is centred around large-volume-market pulp production, but it creates an ecosystem for smaller firms to utilise some of the sidestreams created in the material- and energy-efficient pulp production process.

What do the developments reviewed above imply for the forest-based sector? Consider a simple case, in which the forest-based industries in the USA, Canada, Sweden and Finland gained a 1–2% global market share in the construction, textiles, biofuels, platform chemicals and (plastic) packaging markets by 2030. This could result in an increase in revenue to the forest industries ranging from €18 to 75 billion per annum, corresponding to 10–43% of the production value of the forest industries in these four countries in 2016 (Hurmekoski et al. 2018a). Achieving the higher end of the range would require moving further downstream in the value chains; that is, to assume new roles rather than remain a producer of intermediate goods, which would highlight the role of services.

What would this mean for forests? The impact of gaining a 1–2% market share from primary wood use could be in the range of 15–133 million m^3, corresponding to 2–21% of the industrial roundwood use in these countries in 2016 (Hurmekoski et al. 2018a). The majority of this demand would be for sawn wood for construction. Table 4.3 shows the additional impacts from a 10- and 100-fold market volume. The purpose of such hypothetical scenarios is not to predict the market developments, but simply to assess the scale of the emerging market opportunities. For example, the implications of different market diffusion scenarios can be compared to the net annual increment of forests in the EU (721 million m^3: Forest Europe 2015) or to the global roundwood production (2028 million m^3: FAOSTAT), to make the matter more tangible.

At least two observations arise. Firstly, a minute market share of the global markets would completely transform the forest-based sector. Secondly, it is not realistic to expect an expanding bioeconomy to fully cover the demand in any of these

Table 4.3 Approximate impacts of hypothetical market diffusion scenarios for roundwood and byproduct demand in 2030

~1% market share	Construction	Textiles	Biofuels	Biochemicals	Plastics and packaging	Total
Roundwood, Mm³	7–117	7–15	–	–	2	15–133
Byproducts, Mt	2	–	28	33–45	2	66–73
~10% market share	Construction	Textiles	Biofuels	Biochemicals	Plastics and packaging	Total
Roundwood, Mm³	70–1168	65–147	–	–	15	150–1331
Byproducts, Mt	20	–	280	331–452	25	656–732
~100% market share	Construction	Textiles	Biofuels	Biochemicals	Plastics and packaging	Total
Roundwood, Mm³	698–11,684	650–1469	–	–	153	1501–13,306
Byproducts, Mt	200	–	2800	3311–4520	246	6557–7320

Adapted from Hurmekoski et al. (2018a)

markets, even though the amount of virgin wood resources required to satisfy the construction or textile markets, for example, could, in principle, remain surprisingly small—around the annual increment of the EU forests.

It can be assumed that many of the new products will be based on the existing byproduct flows of the sawmilling and pulping industry, due to the limited ability to pay for the feedstock. Here, wood-based construction is an important driver for raw material availability, both for the pulp and paper industries and for a number of emerging industries, creating both synergies and trade-offs (Hurmekoski et al. 2018a). On one hand, the forest-based product industries would benefit from the increased demand for byproducts (wood chips, bark, sawdust and forest residues), while an increasing production of sawn wood would make generous amounts of byproducts available for the market. On the other hand, there would be competition for the byproducts between the traditional and emerging uses, such as wood-based panels and chemicals or biofuels. Also, other forms of interdependencies between the industries are feasible, such as integrated biofuel and biochemical production in a pulp mill, which could help to lower the pretreatment and transportation costs, and improve energy efficiency (e.g. Kohl et al. 2013; Karvonen et al. 2018).

The future of the forest-based bioeconomy is dependent on the demand from the end markets, developments in substitute markets, biomass markets, as well as policies at varying levels (Hetemäki and Hurmekoski 2020). Due to the variation between and within markets and regions, the opportunities for the forest-based bioeconomy will vary, such that a single, successful strategy cannot be articulated. A greater volume and broader scope of research on these topics would undoubtedly help to enable the emerging opportunities to be grasped.

4.3 Outlook for the Demand for a Forest-Based Bioeconomy

The year 2030 is relatively close, in terms of the likelihood of major new structural changes having impacts in the markets. Thus, a *trend forecast* (using data from 2010 to 2018) to 2030 can provide a helpful baseline against which different assumptions of changes in policy and technology, for example, can be reflected (see Figs. 4.2 and 4.3). In contrast, beyond 2030, major structural changes are likely to occur and have a profound impact on the market, and therefore it is not as helpful to use current trends to provide outlooks that cover several decades (UNECE/FAO 2021).

As the trend projections indicate, the outlook for forest-based products in the coming decade seems increasingly diverse. Information and communications technologies will have several impacts on the demand for forest-based products. These will reduce the demand for communication (graphic) papers, but increase the demand for packaging paper grades, due to the boost in e-commerce (Hetemäki et al. 2013). The demand for consumer papers, such as tissue paper, is expected to continue to grow due to globally increasing middle-income consumers and urbanisation. There are likely to be regional differences in the demand patterns, such as between the OECD and non-OECD countries (Hetemäki et al. 2013). Packaging and tissue-paper consumption are expected to increase, particularly in Asia and Latin America (Pöyry Inc. 2015). Compared to many other industries, the COVID-19 pandemic may have had a relatively small impact on the forest- based industries, but it may alternatively have accelerated the decline in the communication-paper market and the increase in packaging and hygiene paper demand.

Solid wood products have experienced internal competition. The global per-capita sawnwood consumption has declined in a trend-like manner for decades, despite continued growth in the global GDP and population. This is partly explained by the rapidly growing demand for wood-based panels and EWPs that can serve the same markets as sawn wood (Bühlmann and Schuler 2013). For example, the production of CLT in Europe has been growing at an average annual rate of 15% since 2007, despite the economic downturn and stagnation (Pahkasalo et al. 2015). While the overall market share of wood in construction may not necessarily increase significantly by 2030, its growth in hotspots in certain regions and market segments seems likely. The regional differences in demand patterns for solid wood products resemble those for paper products. Notably, China's share of the global wood-based panel production has increased to close to 50% during the last two decades (Fig. 4.1).

The outlook for bioenergy is highly uncertain, and regional differences can be significant (Hetemäki et al. 2020). Of the global roundwood production, roughly half ends up as wood fuel (energy). The major users of wood fuel are Africa, Asia and South America, while Europe's share of the global consumption of wood fuel was 9% (173 million m^3) in 2018 (FAOSTAT). In the long-term, it seems likely that wood-fuel consumption will decline due to a transformation to other energy forms, such as solar, wind, natural gas, hydro and hydrogen, and increases in bioenergy efficiency (Glenn and Florescu 2015; Hetemäki et al. 2020). However, the transformation will also depend on how extensive, and in what time frame, bioenergy carbon capture and storage becomes a viable option. In the nearer future—in the 2020s—and especially in the EU, policies supporting the use of bioenergy may continue to increase the wood-fuel consumption, as major changes in the energy infrastructure tend to take a decade or more, although the likelihood of continued support for the policy remains contested (Hurmekoski et al. 2019).

For emerging forest-based products, their competitiveness depends heavily on the rate of innovation uptake in competing industries, which, in turn, depends on factors such as climate and energy policies and their impacts on oil feedstock and CO_2 emissions prices. For example, synthetic biology, or the conversion of fossil CO2 by industrial biotechnology routes, may have the prospect of providing new products, such as bioethanol and butanol, or organic acids for polyesters, at less cost and using more energy-efficient processes (BIO-TIC 2015; Kruus and Hakala 2017). Carbon dioxide captured from air can technically be directly converted into methanol fuel or plastics (Kothandaraman et al. 2016). Some of these technologies are expected to have already been commercialised by 2030 (BIO-TIC 2015).

It is unclear what the *net impacts* of forest-based product development on global roundwood demand will be (Hetemäki et al. 2020). Some trends point to a growing roundwood demand, including economic and population growth and the need to replace fossil-based materials and energy with more sustainable raw materials (Pepke et al. 2020). For example, Dasos Capital Oy (2019) projected that all the main forest-based products would grow at rates of 2.1–5.5% per annum until 2030. At the same time, the graphic-papers demand is declining and the wood-fuel (bio-energy) demand could decline significantly in Africa and Asia (Hetemäki et al. 2020). Also, resource-efficiency and resource-recovery trends (e.g. cascading, recycling, process technological improvements) may limit the demand for virgin raw materials. Importantly, much of the production of new wood-based products would be based on the byproducts of the mature industries (Hurmekoski et al. 2018a). Determining the extent to which the demand for wood resources for new products would be *additional,* rather than shifting from one use to another, would need to be determined by an optimisation model that also included new wood-based products (see Sect. 4.4). In addition to the challenge to compute net impacts from these diverse trends, the regional differences can be significant. Also, in regions where there is pressure for increasing forest biomass utilisation, there could be trade-offs between the different ecosystem services that forests provide, which could curb the demand.

In summary, through this decade (the 2020s), the trends we are observing today, that are summarised above, will most likely still be the dominant ones. In the longer term, forests will no doubt continue to provide products for the increasing needs of humanity, but the race towards a more sustainable economy will also undoubtedly shape the competition between the forest-based sector and other sectors. The net impacts of the structural changes of the forest-based sector reviewed in the previous sections remain very uncertain. In the next section, we will briefly review the methodological challenges associated with long-term forest-sector outlook studies that partly explain the lack of a more concrete picture of the coming decades.

Box 4.3 Where Will New Jobs Be Created?

In 2018, in the EU27 (excluding the UK), there were 2.1 million people working in the 'traditional' forest sector, comprising forestry, the solid wood industries (excluding furniture) and the pulp and paper industries. Unfortunately, the employment numbers in the sectors such as forest-based bioenergy, biochemicals, biotextiles, furniture, printing, etc. cannot be derived from existing statistical classifications, However, together, they could be even more than the employment in the traditional forest sector. The furniture industry alone employed 1.13 million in 2018, and probably a significant portion of that was based on wood furniture. Thus, it would not be surprising if all the forest-based industries with value-added activities in the EU27 had employed around four million people. Moreover, there are 16 million private forest owners in the EU, many of whom earn income from forests.

Given that fossil-based production needs to be phased out, this will inevitably mean that many people will need to find new jobs in the EU. For example, very basic, fossil-based raw-material manufacturing (coke and refined petroleum products) alone employed a total of 186,200 people in 2018 in the EU27. If we included the fossil-based-plastics industry and other fossil-based industry sectors, this number would be much greater. For example, rubber and plastics manufacturing alone employed 110,400 persons in 2018 in the EU27.

We are not aware of any projections for how employment in the fossil-based manufacturing sector will develop in the EU27 when we move towards carbon neutrality by 2050. In order to have at least some idea of the importance of the question, let us review a hypothetical example. Assume that CO_2 emissions from the manufacture of coke, crude oil, gas and petroleum in the EU is phased out from 2020 to 2050 by 2.5% points each year. This would imply that, by 2050, there would be a cut of 75% of this sector's CO_2 emissions. For the sake of simplicity, let us also assume that employment in the sector is falling at the same rate; that is, only 25% of the employment level of 2020 is left in 2050. That would mean the loss of about 140,000 jobs in the fossil sector of the EU27 during this period (Fig. Box 4.2). Naturally, the rate could be different, and there would be other fossil-based sectors, such as plastics and chemicals, also losing jobs.

The European Green Deal (EGD) acknowledges the need for a socially justifiable move towards carbon neutrality by 2050, for example, via compensation funds to countries and regions that are heavily dependent on fossil-based industries. However, these funds will not automatically and necessarily result in new jobs. Therefore, it will be crucial for the EGD to address more explicitly how new jobs will be created, and in which sectors, when we move towards a carbon-neutral society.

(continued)

Box 4.3 (continued)

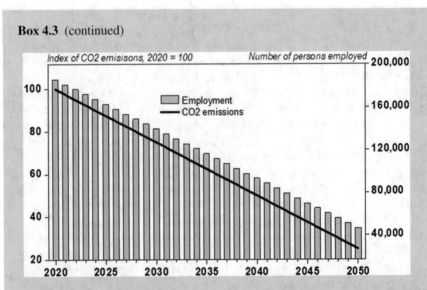

Fig. Box 4.2 Illustrative example of CO$_2$ emissions and number of people employed in the EU27 in coke, crude oil, gas and petroleum manufacturing in 2020–2050. (Source of employment data for 2018: EUROSTAT)

The creation of new jobs becomes an ever-bigger challenge if the EGD imposes policies that work towards cutting back economic activities in the renewable biological-resource-based sectors. Rather, the objective should be to boost the biological sector's economic activities in order to help phase out the fossil sectors, and to create employment opportunities in more-sustainable economic sectors. But this, of course, has to be done in an even more sustainable and resource-efficient way than in the past, while also considering the biodiversity needs. To this end, the approach outlined by researchers, termed 'climate-smart forestry', could provide the way forward in the forest sector (see Chap. 9).

4.4 Research Implications

As pointed out throughout this chapter, research on the future of the forest-based sector remains scarce. Moreover, the focus of outlook studies has primarily been on the sufficiency of wood resources, trends in the production of primary forest-based products and international competitiveness, while some of the equally important questions, such as value added, employment, changing demand patterns, and emerging forest-based products and services, have gained less attention, until recently (Hetemäki and Hurmekoski 2016; UNECE/FAO 2021).

The mainstream of the forest-sector outlook studies has been based on forest-sector modelling, combined with stakeholder interaction and scenario development, as a way of exploring and addressing the possible trade-offs that decision-makers will face (Hurmekoski and Hetemäki 2013). The strength of the forest-sector models is in capturing the interdependencies between different parts of the system, modelling the feedback and trickle-down impacts of changes in one part of the sector on the rest of the sector by making market adjustments through pricing and international trade (e.g. Toppinen and Kuuluvainen 2010).

The drawback of partial-equilibrium modelling and traditional econometrics is that determining the impacts of an exogenous shock requires a stable operating environment and the use of historical data, usually from a period of several decades. However, we have observed several major structural changes in the global and European forest-based sector in this century, which past data, and models based on these, have difficulties in capturing. The structural changes are a result of two primary factors—the expected market diffusion of new wood-based products and the changes in demand patterns for some of the established product groups. Regarding new wood-based products, the apparent limitation is the lack of reliable data (or long enough time-series) for the purpose of traditional econometric analysis. The changing demand patterns may be a less apparent, albeit equally important, factor to consider in modelling. However, apart from single cases, such as the substitution of graphic papers by electronic media (Hetemäki and Obersteiner 2001; Latta et al. 2016), the demand shifters for most forest-based products remain elusive, so that incorporating the omitted variables in demand equations may not necessarily be accomplished in a completely satisfactory way.

Thus, the strength of the forest-sector models in capturing the interdependencies between different parts of the forest sector can turn into a weakness in the presence of structural changes in demand. It can lead to distorted feedback effects, which may compromise the internal logic of the models. This means that, even if the primary research question was unrelated to structural changes, their impact may distort the findings in any 'what if' analysis or long-term projection. For example, we cannot assess whether an increase in forest-products production in one region, *ceteris paribus*, will signify a decrease in wood-products production in some other region, or a decrease in production of other materials. In the event of changes in the intersectoral market share, we are also lacking the means to reliably assess changes in the international market shares, even though it would otherwise be quite straightforward. This is a significant drawback when assessing the future uses of wood and their implications on the economy and the environment.

Models are always simplifications of reality and can never be expected to accurately predict market developments for several decades ahead. However, since forest-products-markets research clearly falls into the realm of applied science, it is essential to try to capture and explain market developments in order to maintain their practical relevance (Hetemäki and Hurmekoski 2016). Although the evidence-based models continue to be crucial elements of forest-sector outlook studies, they are unlikely to meet the needs of decision-making alone in the increasingly complex forest-based sector (Toppinen and Kuuluvainen 2010; Hurmekoski and Hetemäki 2013). For some forest products or regions, the traditional modelling framework could work well, but to complement the picture with emerging and declining markets, alternative approaches may be necessary.

4.5   Key Messages

- The state of the world's managed forests is determined by the societal demands for wood resources and other ecosystem services. The interplay of supply and demand thereby determines the employment and revenues created by the sector, as well as the ability of forests to provide a range of ecosystem services, such as wood-based products or carbon sinks.
- For the sake of the efficient planning of strategies, it can be useful to look ahead and try to anticipate possible changes that may take place in the future. Indeed, outlook studies have a long tradition in the forest-based sector.However, in the face of increasing structural changes taking place in the sector and the operating environment, the task of looking ahead is becoming evermore important, but also more challenging.
- In the twenty-first century, several structural changes have become evident. The demand for some of the traditional forest products is no longer developing on a par with consumers' available incomes. Instead, there are new factors influencing the demand for forest products, causing substitution between forest products and alternative products. As these factors are partly unknown or unmeasurable, future projections are subject to increasing uncertainty.
- A key trend in the forest industries is diversification, referring to new market opportunities for wood-based industries in, for example, the construction, textiles, packaging, biochemical and biofuel markets. As the importance of some of the traditional forest products—notably communication papers—are declining, the sector will not necessarily be dominated by single sectors in the long term. Moreover, the boundaries with other industries are becoming increasingly indistinguishable.
- There is a severe lack of systematic outlook studies that would illuminate the possible impacts of the expected structural changes of the forest-based sector (Hetemäki et al. 2020). As changes in the forest sector are usually gradual and slow, compared to the digital sector, for example, the current trends are likely set in a reasonable direction up to around 2030. However, over time, the uncertainties will grow bigger, owing to the large number of emerging market opportunities, as well as developments in the operating environment and other sectors. These uncertainties also imply significant uncertainties in the global demand for roundwood, and by extension, also the extent of the trade-offs between different ecosystem services and land uses. The implications for the climate impacts of wood use are reflected further in Chap. 7.

Acknowledgements Elias Hurmekoski wishes to acknowledge financial support from the SubWood Project (No. 321627), funded by the Academy of Finland.

References

Aeschelmann F, Carus M (2015) Bio-based building blocks and polymers in the world – capacities, production and applications: status quo and trends towards 2020. Summary. Nova-Institut für politische und ökologische Innovation Gmbh, Hürth

Alderman D (2013) Housing and construction markets. UNECE/FAO Forest products annual market review 2012–2013. Geneva timber and Forest study paper 33. Forestry and Timber Section, United Nations, Geneva, pp 115–122

Alkhagen M, Samuelsson Å, Aldaeus F, Gimåker M, Östmark E, Swerin A (2015) Roadmap 2015 to 2025. Textile materials from cellulose. Research Institutes of Sweden

Alper E, Orhan OY (2017) CO_2 utilization: developments in conversion processes. Petroleum 3:109–126

Anderson P, Tushman ML (1990) Technological discontinuities and dominant designs: a cyclical model of technological change. Adm Sci Q 35:604–633

Antikainen R, Dalhammar C, Hildén M, Judl J, Jääskeläinen T, Kautto P, Koskela S, Kuisma M, Lazarevic D, Mäenpää I (2017) Renewal of forest based manufacturing towards a sustainable circular bioeconomy. Reports of the Finnish Environment Institute 13

Aro T, Fatehi P (2017) Tall oil production from black liquor: challenges and opportunities. Sep Purif Technol 175:469–480

Arora SK, Foley RW, Youtie J, Shapira P, Wiek A (2014) Drivers of technology adoption—the case of nanomaterials in building construction. Technol Forecast Soc Change 87:232–244

Arrow KJ (1962) The economic implications of learning by doing. Rev Econ Stud 29:155–173

Bazzanella AM, Ausfelder F (2017) Low carbon energy and feedstock for the European chemical industry. Technology Study. DECHEMA Gesellschaft für Chemische Technik und Biotechnologie e.V., Study commissioned by Cefic, the European Chemical Industry Council

Bell W (2003) Foundations of futures studies: History, purposes, and knowledge. Transaction Publishers, Piscataway

Biddy MJ, Scarlata C, Kinchin C (2016) Chemicals from biomass: a market assessment of bio-products with near-term potential. National Renewable Energy Laboratory, Golde. https://www.nrel.gov/docs/fy16osti/65509.pdf

Bio-based Industries Consortium (2013) The Strategic Innovation and Research Agenda (SIRA). https://biconsortium.eu/about/our-vision-strategy/sira. Accessed 13 Mar 2020

BIO-TIC (2015) A roadmap to a thriving industrial biotechnology sector in Europe. http://www.industrialbiotech-europe.eu/wp-content/uploads/2015/08/BIO-TIC-roadmap.pdf. Accessed 30 March 2020

Bruijnincx P, Weckhuysen B, Gruter G-J, Westenbroek A, Engelen-Smeets E (2016) Lignin valorization. The importance of a full value chain approach. Utrecht University, Utrecht

Bühlmann U, Schuler A (2013) Markets and market forces for secondary wood products. In: Hansen E, Panwar R, Vlovsky R (eds) The global Forest sector. Changes, practices, and Prospects. CRC Press, Boca Raton

Byun Y, Kim YT (2014) Utilization of bioplastics for food packaging industry. In: Han JH (ed) Innovations in food packaging, 2nd edn. Academic Press/Elsevier, Amsterdam

Cai Z, Rudie AW, Stark NM, Sabo RC, Ralph SA (2013) New products and product categories in the global forest sector. In: Hansen E, Panwar R, Vlosky R (eds) The global forest sector. Changes, practices. and Prospects. CRC Press, Boca Raton, pp 129–150

Carus M, Eder A, Dammer L, Korte H, Scholz L, Essel R, Breitmayer E, Barth M (2015) Wood –plastic composites (WPC) and natural fibre composites (NFC): European and global markets 2012 and future trends in automotive and construction. Nova-Institut für politische und ökologische Innovation GmbH, Hürth

Carus M, Raschka A, Iffland K, Dammer L, Essel R, Piotrowski S (2016) How to shape the next level of the European bio-based economy. The Reasons for the Delay and the Prospects of Recovery in Europe. Renewable matter. NOVA Institute

Cowie J, Bilek EM, Wegner TH, Shatkin JA (2014) Market projections of cellulose nanomaterial-enabled products—part 2: volume estimates. TAPPI J 13:57–69

Dasos Capital (2019) Future prospects for forest products and timberland investment, 5th edn. Dasos Capital Oy, Helsinki de Jong E, Higson A, Walsh P, Wellisch M (2012) Bio-based chemicals value added products from biorefineries. Task 42 Biorefineries. International Energy Agency Bioenergy

de Jong E, Higson A, Walsh P, Wellisch M (2012) Bio-based chemicals value added products from biorefineries. IEA Bioenergy, Task 42 Biorefinery

de Vet J-M, Pauer A, Merkus E, Baker P, Gonzalez-Martinez AR, Kiss-Galfalvi T, Streicher G, Rincon-Aznar A (2018) Competitiveness of the European cement and lime sectors. European Commission, Brussels

Dornburg V, Hermann BG, Patel MK (2008) Scenario projections for future market potentials of biobased bulk chemicals. Environmental Science and Technology 42:2261–2267

Edita Prima Ltd, Helsinki. Textile Exchange (2019) Preferred fiber & materials. Market Report 2019. Textile Exchange

European Commission (2018) A sustainable bioeconomy for Europe – strengthening the connection between economy, society and the environment: updated bioeconomy strategy. European Commission, Brussels

European Union (2009) Directive 2009/28/EC of the European Parliament and of the council of 23 April 2009 on the promotion of the use of energy from renewable sources and amending and subsequently repealing Directives 2001/77/EC and 2003/30/EC

FitzPatrick M, Champagne P, Cunningham MF, Whitney RA (2010) A biorefinery processing perspective: treatment of lignocellulosic materials for the production of value-added products. Bioresource Technology 101:8915–8922

Food and Agriculture Organization (2018) The State of the World's Forests 2018 – Forest pathways to sustainable development. Licence: CC BY-NC-SA 3.0 IGO. FAO, Rome

Food and Agriculture Organization, Rome (2016) Forestry for a low-carbon future: integrating forests and wood products into climate change strategies. FAO, Rome

Forest Europe (2015) State of Europe's forests 2015. Forest Europe, Bonn

Glenn JC, Florescu E (2015) 2015–2016 state of the future. The Millenium Project, Washington, DC

Graichen FHM, Grigsby WJ, Hill SJ, Raymond LG, Sanglard M, Smith DA, Thorlby GJ, Torr KM, Warnes JM (2016) Yes, we can make money out of lignin and other bio-based resources. Ind Crop Prod 106. https://doi.org/10.1016/j.indcrop.2016.10.036

Guerrero JE, Hansen E (2018) Cross-sector collaboration in the forest products industry: a review of the literature. Can J For Res 48:1269–1278

Hämäläinen S, Näyhä A, Pesonen H-L (2011) Forest biorefineries – a business opportunity for the Finnish forest cluster. J Clean Prod 19:1884–1891

Hammerle FM (2011) The cellulose gap (the future of cellulose fibers). Lenzinger Ber 89:12–21

Hetemäki L (1999) Information technology and paper demand scenarios. In: Palo M, Uusivuori J (eds) World forests, society and environment. Springer, Dordrecht, pp 31–40

Hetemäki L (2005) ICT and communication paper markets. Inf Technol For Sect 18:76–104

Hetemäki L, Hurmekoski E (2016) Forest products markets under change: review and research implications. Curr For Rep 2:177–188. https://doi.org/10.1007/s40725-016-0042-z

Hetemäki L, Hurmekoski E (2020) Forest bioeconomy development: markets and industry structures. In: Nikolakis W, Innes JL (eds) The wicked problem of forest policy: multidisciplinary approach to sustainability in forest landscapes. Cambridge University Press, Cambridge

Hetemäki L, Obersteiner M (2001) US newsprint demand forecasts to 2020. International Institute for Applied Systems Analysis (IIASA), Interim Report IR-01-070. International Institute for Applied Systems Analysis, Laxenburg

Hetemäki L, Hänninen R, Moiseyev A (2013) Markets and market forces for pulp and paper products. In: Hansen E, Vlosky R, Panwar R (eds) Global Forest products: trends, management, and sustainability. Taylor and Francis Publishers, Abingdon

Hetemäki L, Palahi M, Nasi R (2020) Seeing the wood in the forests. Connecting knowledge to action no.1. European Forest Institute, Joensuu

Hildebrandt J, Hagemann N, Thrän D (2017) The contribution of wood-based construction materials for leveraging a low carbon building sector in Europe. Sustain Cities Soc 34:405–418

Hurmekoski E, Hetemäki L (2013) Studying the future of the forest sector: review and implications for long-term outlook studies. For Policy Econ 34:17–29

Hurmekoski E, Jonsson R, Korhonen J, Jänis J, Mäkinen M, Leskinen P, Hetemäki L (2018a) Diversification of the forest- based sector: role of new products. Can J For Res 48:1417–1432. https://doi.org/10.1139/cjfr-2018-0116

Hurmekoski E, Pykäläinen J, Hetemäki L (2018b) Long-term targets for green building: explorative Delphi back casting study on wood-frame multi-story construction in Finland. J Clean Prod 172:3644–3654. https://doi.org/10.1016/j.jclepro.2017.08.031

Hurmekoski E, Lovrić M, Lovrić N, Hetemäki L, Winkel G (2019) Frontiers of the forest-based bioeconomy–a European Delphi study. For Policy Econ 102:86–99

Jonsson R, Hurmekoski E, Hetemäki L, Prestemon J (2017) What is the current state of forest product markets and how will they develop in the future? In: Winkel G (ed) Towards a sustainable European forest-based bioeconomy – assessment and the way forward. What science can tell us. European Forest Institute, Joensuu, pp 126–131

Kangas K, Baudin A (2003) Modelling and projections of forest products demand, supply and trade in Europe. A study prepared for the European Forest Sector Outlook Study (EFSOS). Food and Agriculture Organization, Rome

Karvonen J, Kunttu J, Suominen T, Kangas J, Leskinen P, Judl J (2018) Integrating fast pyrolysis reactor with combined heat and power plant improves environmental and energy efficiency in bio-oil production. J Clean Prod 183:143–152

Kemppainen K, Siika-aho M, Pattathil S, Giovando S, Kruus K (2014) Spruce bark as an industrial source of condensed tannins and non-cellulosic sugars. Ind Crop Prod 52:158–168

Klemperer WD (2003) Forest resource economics and finance. McGraw-Hill, New York

Kohl T, Laukkanen T, Järvinen M, Fogelholm C-J (2013) Energetic and environmental performance of three biomass upgrading processes integrated with a CHP plant. Appl Energy 107:124–134

Korhonen J, Koskivaara A, Toppinen A (2020) Riding a Trojan horse? Future pathways of the fiber -based packaging industry in the bioeconomy. For Policy Econ 110:101799

Kothandaraman J, Goeppert A, Czaun M, Olah GA, Prakash GKS (2016) Conversion of CO_2 from air into methanol using a polyamine and a homogeneous ruthenium catalyst. J Am Chem Soc 138:778–781

Krasutsky PA (2006) Birch bark research and development. Nat Prod Rep 23:919–942

Kruus K, Hakala T (2017) The making of bioeconomy transformation. VTT Technical Research Centre of Finland Ltd, Espoo

Lähtinen K, Harju C, Toppinen A (2019) Consumers' perceptions on the properties of wood affecting their willingness to live in and prejudices against houses made of timber. Wood Mater Sci Eng 14:325–331

Latta GS, Plantinga AJ, Sloggy MR (2016) The effects of internet use on global demand for paper products. J For 114(4):433–440

Laurichesse S, Avérous L (2014) Chemical modification of lignins: towards biobased polymers. Prog Polym Sci 39:1266–1290

Mahapatra K, Gustavsson L (2008) Multi-storey timber buildings: breaking industry path dependency. Build Res Inf 36:638–648

Malmgren L (2014) Industrialized construction – explorations of current practice and opportunities. Doctoral thesis, Lund University

Mantau U, Saal U, Prins K, Steierer F, Lindner M, Verkerk H, Eggers J, Leek N, Oldenburger J, Asikainen A (2010) EuWood – real potential for changes in growth and use of EU forests. European Forest Institute, Sarjanr

Moiseyev A, Solberg B, Kallio AMI (2013) Wood biomass use for energy in Europe under different assumptions of coal, gas and CO2 emission prices and market conditions. J For Econ 19:432–449

Muilu-Mäkelä R, Haavisto M, Uusitalo J (2014) Puumateriaalien terveysvaikutukset sisäkäytössä – Kirjallisuuskatsaus [Health effects of wooden interiors – a literature review]. Metla working papers 320. Vantaa, Finland

Nägele H, Pfitzer J, Nägele E, Inone ER, Eisenreich N, Eckl W, Eyerer P (2002) ARBOFORM® – a thermoplastic, processable material from lignin and natural fibers. Chemical Modification, Properties, and Usage of Lignin Springer, pp. 101–119

Naik SN, Goud VV, Rout PK, Dalai AK (2010) Production of first and second generation biofuels: a comprehensive review. Renew Sust Energ Rev 14:578–597

Näyhä A, Pesonen H-L (2012) Diffusion of forest biorefineries in Scandinavia and North America. Technol Forecast Soc Change 79:1111–1120

Näyhä A, Hetemäki L, Stern T (2014) New products outlook. In: Hetemäki L (ed) Future of the European forest -based sector: structural changes towards bioeconomy. What science can tell us 6. European Forest Institute, pp 15–32

Neil-Boss N, Brooks K (2013) Unwrapping the packaging industry. Seven Factors for Success. Ernst & Young

Nguyen Q, Bowyer J, Howe J, Bratkovich S, Groot H, Pepke E, Fernholz K (2017) Global production of second generation biofuels: trends and influences. Dovetail Partners Inc

Pahkasalo T, Gaston C, Schickhofer G (2015) Value-added wood products. In: UNECE/FAO (ed) Forest products annual market review 2014–2015. Switzerland, Geneva, pp 105–114

Textile Exchange (2019) Preferred Fiber & Materials Market Report 2019

Pepke E, Bowyer J, Fernholz K, Groot H, McFarland A (2020) Future developments in the forest sector. Dovetail Partners Inc. Plastics Europe (2016) Plastics – the Facts 2016. An analysis of European plastics production, demand and waste data. Brussels, Belgium

Pöyry Inc. (2015) World fibre outlook up to 2030. Vantaa, Finland

Rosenberg N (1982) Inside the black box: technology and economics. Cambridge University Press, Cambridge

Rougieux P, Damette O (2018) Reassessing forest products demand functions in Europe using a panel cointegration approach. Appl Econ 50(30):3247–3270

Routa J, Brännström H, Anttila P, Mäkinen M, Jänis J, Asikainen A (2017) Wood extractives of Finnish pine, spruce and birch–availability and optimal sources of compounds: a literature review. Nat Resour Bioeconomy Stud 73:55

Routley M, Phaal R, Probert D (2013) Exploring industry dynamics and interactions. Technol Forecast Soc Change 80:1147–1161

Schipfer F, Kranzl L, Leclère D, Sylvain L, Forsell N, Valin H (2017) Advanced biomaterials scenarios for the EU28 up to 2050 and their respective biomass demand. Biomass Bioenergy 96:19–27

Shen L, Worrell E, Patel MK (2010) Environmental impact assessment of man-made cellulose fibres. Resour Conserv Recycl 55:260–274

Shirmohammadli Y, Efhamisisi D, Pizzi A (2018) Tannins as a sustainable raw material for green chemistry: a review. Ind Crop Prod 126:316–332

Sirajunnisa AR, Surendhiran D (2016) Algae–a quintessential and positive resource of bioethanol production: a comprehensive review. Renew Sust Energ Rev 66:248–267

Soroudi A, Jakubowicz I (2013) Recycling of bioplastics, their blends and biocomposites: a review. Eur Polym J 49:2839–2858

Stern T, Ledl C, Braun M, Hesser F, Schwarzbauer P (2015) Biorefineries' impacts on the Austrian forest sector: a system dynamics approach. Technol Forecast Soc Change 91:311–326

Taylor R, Nattrass L, Alberts G, Robson P, Chudziak C, Bauen A, Libelli IM, Lotti G, Prussi M, Nistri R (2015) From the sugar platform to biofuels and biochemicals. Final Report for the European Commission Directorate-General Energy N (ENER/C2/423-2012/SI2.673791)

ministry of Employment and the Economy (2014) Sustainable growth from bioeconomy. The Finnish Bioeconomy Strategy

Thomas B, Raj MC, Joy J, Moores A, Drisko GL, Sanchez C (2018) Nanocellulose, a versatile green platform: from biosources to materials and their applications. Chem Rev 118:11575–11625

Toppinen A, Kuuluvainen J (2010) Forest sector modelling in Europe—the state of the art and future research directions. For Policy Econ 12(1):2–8. https://doi.org/10.1016/j.forpol.2009.09.017

Toppinen A, Röhr A, Pätäri S, Lähtinen K, Toivonen R (2018) The future of wooden multistory construction in the forest bioeconomy – a Delphi study from Finland and Sweden. J For Econ 31:3–10

United Nations (2015a) Transforming our world: The 2030 agenda for sustainable development. United Nations General Assembly, A/RES/70/1

United Nations (2015b) Paris agreement. United Nations

United Nations Economic Commission for Europe/Food and Agriculture Organization (2011) The European forest sector outlook study II 2010–2030. UNECE/FAO, Geneva

United Nations Economic Commission for Europe/Food and Agriculture Organization (2021) Forest sector outlook study III. UNECE/FAO, Geneva

Uronen T (2010) On the transformation processes of the global pulp and paper industry and their implications for corporate strategies – a European perspective. Doctoral dissertation, Aalto University

Varian H (2010) Intermediate microeconomics, 8th edn. WW Norton & Company, Inc, New York

World Economic Forum (2016) The new plastics economy - rethinking the future of plastics

Chapter 5
Forest Biomass Availability

Perttu Anttila and Hans Verkerk

Abstract The forest-based bioeconomy relies on using forests as a source of raw material for producing materials and energy, as well as for a variety of other ecosystem services. The uses of forests and wood are many and, to some extent, competing. Can a limited resource simultaneously and sustainably provide raw materials for products, feedstock for energy production, and other ecosystem services? Over one-third of the land area in the EU is covered by forests, but there are large differences between the member states regarding both forest area and growing stock of wood. The harvesting of roundwood has been steadily increasing. In addition to roundwood, other tree parts, as well as residues from forest industries and post-consumer wood, are being used for both materials and energy production. There are non-negligible uncertainties regarding the future availability of forest biomass in the context of climate change, as well as difficulties to concern all the relevant constraints on biomass supply in relation to availability assessments and the difficult-to-predict effects of policies. Despite the above, it can be concluded that there is still potential to increase the utilisation of forest biomass in most of the EU regions, but this might affect the provisioning of other important ecosystem services.

Keywords Forest resources · Harvesting potential · Roundwood · Forestry residues · Utilisation rate

P. Anttila (✉)
Natural Resources Institute Finland (Luke), Helsinki, Finland
e-mail: perttu.anttila@luke.fi

H. Verkerk
European Forest Institute, Joensuu, Finland

5.1   European Forests and the Utilisation of Biomass

Aside from other services, the EU's forests represent a vast raw-material resource. Forests account for 38% of the EU28 land area (Forest Europe 2020a). In 2019, this amounted to 162 mill. ha in total, of which 138 mill. ha was available for wood supply. Forests are, however, unevenly distributed, and the share of forest area is generally higher in Northern Europe (up to 74% in Finland) than in Central or Southern Europe, where it is between 30 and 40% in many countries (e.g. France, Germany, Italy, Poland and Spain), or even less. Of the total EU28 forest area, 60% occurs in only five member states—Sweden, Finland, Spain, France and Germany.

The growing stock of the EU28 forests available for wood supply amounts to nearly 23 billion m^3 (Forest Europe 2020a). The five biggest growing-stock countries account for 60% of this total, these being Germany (3.5 billion m^3), France (2.9 billion m^3), Sweden (2.7 billion m^3), Poland (2.4 billion m^3) and Finland (2.2 billion m^3). The high volumes in Central Europe can be explained by a high stocking density (m^3 ha^{-1}) in the forests in relation to the somewhat lower density in Northern Europe. The above figures considered stemwood only, but branches and stumps are also potential sources of woody biomass. Such woody components can increase the aboveground biomass by 50% (Camia et al. 2018).

The woody biomass used to produce materials and energy comes from various sources. In 2015, the share of woody biomass in the EU28 that was harvested directly from domestic forests was nearly 57%, and the rest originated from imports, byproduct and coproduct supply, wood pellet supply, post-consumer wood and unaccounted sources (Cazzaniga et al. 2019b). Forest biomass comprises roundwood and primary residues, i.e., logging residues (consisting of crown biomass and stemwood loss), small-diameter trees and stumps.

There has been a clearly increasing trend in roundwood harvesting volumes in the EU28 in this millennium, apart the financial crisis that caused a slump in 2008–2009 (see Fig. 1.4 in Chap. 1 – Box: Forest Bioeconomy in the EU). Roundwood production increased from 486 mill. m^3 (overbark) in 2000 to 578 mill. m^3 in 2019—an increase of almost one-fifth (Food and Agriculture Organization [FAO] 2020). The five countries with the largest forest resources for wood production also harvest most of the roundwood (Fig. 5.1). In Fig. 5.1, the impact of the economic slump due to the financial crisis in 2009 is clearly visible.

Several severe storms have occurred in recent years in Europe (Forzieri et al. 2020; Senf and Seidl 2020) and their effects can be seen in terms of harvested volumes. For example, storm Gudrun in January 2005 damaged 75 mill. m^3 of roundwood in Sweden, causing a supply peak of more than 30 mill. m^3 (Gardiner et al. 2010). Likewise, the peak caused by storm Kyrill in January 2007, which damaged 37 mill. m^3, is also discernible in Fig. 5.1. However, contrary to the two abovementioned storms, the effect of storm Klaus in February 2009 in France, which damaged 43 mill. m^3, is barely visible. Similarly, storms and other disturbances in different countries have also had significant effects. For example, the storm Vaia in Italy resulted in about 8 mill. m^3 of damaged wood— approximately the same

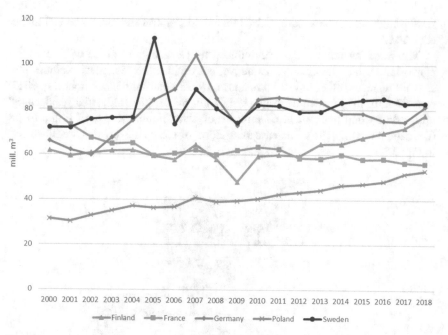

Fig. 5.1 Harvesting volumes (overbark) of roundwood in five EU member states in 2000–2018 (FAO 2020). Underbark figures converted to overbark using the coefficient 1/0.88 (FAO, International Tropical Timber Organization and United Nations 2020)

amount as harvested in an entire year in Italy. Also, the recent dry summers in Central Europe, followed by severe bark-beetle outbreaks, have produced large amounts of damaged wood. Disturbances can have a strong impact on the local forest sector, first by creating a pulse of available timber from salvage harvesting, but later resulting in a shortage of local timber supply.

In addition to roundwood, primary residues are also being utilised, but mainly in energy production. Unfortunately, there are no EU-level statistics on the consumption of primary residues, and even national statistics may be weak. Germany, Sweden and Finland are probably the top three countries in the EU. According to Brosowski et al. (2016), the annual consumption of logging residues for energy use in Germany in 2012 was 4.0–10.5 Tg. Assuming a basic density of 400 kg m^{-3} for conifers and 500 kg m^{-3} for broadleaves, the consumption would have been roughly 9–23 mill. m^3. In Sweden, the consumption of forest fuels between 2013 and 2018 was 15.2–20.2 TWh (Energimyndigheten 2020). This equates to approximately 8–10 mill. m^3, assuming 1 solid m^3 equals 2 MWh. In Finland, the consumption for the same period was 7–8 mill. m^3 (Natural Resources Institute 2020).

The other sources of woody biomass can be divided into secondary and tertiary forestry residues and trade. Trees grown outside of forests and short-rotation coppice grown on forest or agricultural land are minor sources that are not discussed here. The secondary forestry residues (aka industrial residues) are the side products of wood processing or come from the production of wood products, and include sawdust and cutter chips, bark, slabs, lumpwood residues and black liquor (Lindner et al. 2017).

The wood resource balance introduced by Cazzaniga et al. (2019b) does not employ the same classification as above, but divides the secondary residues into sawmill residues, other industrial residues, wood pellets and black liquor. In 2015, over 87 mill. m³ of sawmill residues, 11 mill. m³ of other industrial residues, 38 mill. m³ of wood pellets and 67 mill. m³ of black liquor were used in the EU28 (Cazzaniga et al. 2019b). The cascading flows of these side-streams are illustrated in Fig. 5.2.

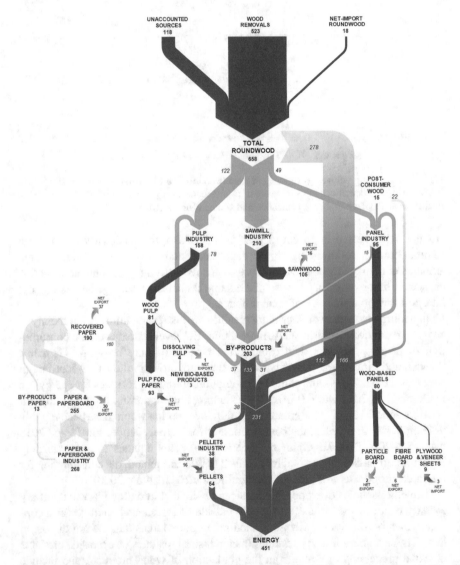

Fig. 5.2 Woody biomass flows in the EU28 in 2015 (in mill. m³ solid wood equivalent overbark). (Source: Cazzaniga et al. 2019a)

Tertiary forestry residues (i.e. post-consumer wood) include wooden material that is available at the end of its use as a wooden product. In 2015, the consumption of tertiary residues in the EU28 was estimated at 37 mill. m³ (Cazzaniga et al. 2019b).

Furthermore, wood is traded between the EU and other countries. Between 2010 and 2018, the EU was a net importer of roundwood, bringing in 11–18 mill. m³ (overbark) more than it exported (FAO 2020). In 2019, however, the direction of the stream reversed, with the EU exporting 2 mill. m³ more than it imported. To a large extent, this amonnt can be attributed to the Central European bark-beetle infestation, which has forced the Czech Republic, in particular, to greatly increase their timber exports.

One single country stands out among the exporters to the EU-Russia alone has exported 7–10 mill. m³ of industrial ronndwood and wood chips between 2010 and 2017 (FAO 2020). Russia's forest area is five times larger than the EU's, and its exports to conntries like Finland and Sweden are substantial. However, basing the feedstock sourcing of bio-based businesses solely on Russian wood imports would be challenging for several political and infrastructural reasons (Leskinen et al. 2020; Box: Huge Russian forest resources -a reality or an illusion?).

5.2 Availability of Forest Biomass

The growing stock and increment rates of Europe's forests have been increasing almost continuously over the last several decades (Gold et al. 2006; Forest Europe 2020a). In fact, Albania is the only country in the whole of Europe that has reported a decrease in growing stock between 1990 and 2015. In recent years, the increase has been especially rapid in Central-East Europe (including Ukraine, Belarus and Georgia outside of the EU).

The major reasons for the increasing growing stock include: the fellings and natural losses that together have been less than the gross increment; the increasing increment rates and changes in forest management that have caused forests to become denser (e.g. Vilén et al. 2016); nitrogen deposition (e.g. de Vries et al. 2009; Etzold et al. 2020); as well as the combined effect of nitrogen deposition, increased atmospheric CO_2 concentrations and climate change (Pretzsch et al. 2014; Flechard et al. 2020).

The relation of annual fellings to the net annual increment (NAI) is a key sustainability indicator of wood production. Generally, if the fellings fall below the NAI, the growing stock is increasing. Correspondingly, if the fellings are more than the NAI, the growing stock is decreasing. On average, 75% of the NAI was utilised by the EU28 in 2015 (Forest Europe 2020a). However, the utilisation rates varied considerably, from 99% in Belgium to 44% in Romania (and probably even lower in countries lacking data).

The NAI is only a rough estimate of the maximum potential availability of wood from forests, as it does not consider the stocking level of the forests, imbalances in forest age structures, the potential availability of biomass from primary residues, or ecological and socioeconomic factors. Furthermore, numerous technical, environmental, economic and social constraints, which limit the availability of forests to harvesting, need to be considered. Such factors can include soil productivity, soil and water protection, biodiversity protection, technical recovery rates, the soil bearing capacity, forest-owner behaviour, the profitability of wood production (harvesting), and regional land-use plans (Verkerk et al. 2011, 2019; Barreiro et al. 2017; di Fulvio et al. 2016; Kärkkäinen et al. 2020). When taking age-structure and stocking level into account, and correcting for the constraints, the woody biomass potential from EU forests has been estimated to range between 663 and 795 mill. m^3 a^{-1}, of which some 80–90% is stemwood, the rest being mainly logging residues (Verkerk et al. 2011, 2019; di Fulvio et al. 2016; Jonsson et al. 2018). These biomass potentials are fairly stable over time.

The highest potentials per unit of land area can be found in parts of Northern Europe (southern Finland and Sweden, Estonia and Latvia), Central Europe (Austria, the Czech Republic and southern Germany), southwest France and Portugal (Fig. 5.3).

Comparing the potential in the EU with the average roundwood production in 2010–2019, which was 539 mill. m^3 (FAO 2020), reveals that over 80% of the stemwood potential is already in use. In fact, the share of utilised potential could be even higher. The number for roundwood production has been found to underestimate fellings due to, for example, unregistered fuelwood fellings in private forests (Jochem et al. 2015). In some of the areas, wood use is already at a high level, indicating little potential for increased use (Verkerk et al. 2019). Such areas include southern Sweden and southwestern France. However, in some regions, the potential could allow a considerable increase in utilisation.

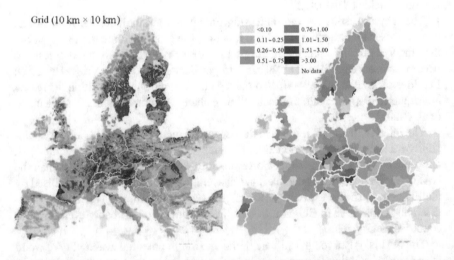

Fig. 5.3 Estimated spatial distribution of forest biomass availability in 2020 (t·ha^{-1}a^{-1}) (left) and unused potential per unit of land (t·ha^{-1}a^{-1}) (right). (Source: Verkerk et al. 2019)

The utilisation rate of primary forestry residues is substantially lower than for stemwood. Yet in places where the utilisation rate is high, the competition from residues can increase supply costs. For example, in southern Finland, the consumption of residues is expected to top the harvesting potential (Anttila et al. 2018).

Dees et al. (2017) estimated the potentials for secondary forestry residues in the EU28 to be 190 and 194 mill. m³ for 2020 and 2030, respectively. The total potential was further broken down into residues from the sawmill industry, pulp and paper industry, and other wood-processing industries, with shares of 43, 37 and 20% in 2012, respectively. Naturally, these potentials depend directly on the production of the industries.

5.3   Outlook for Forest Biomass Availability

Increasing harvesting to the limit of the potentially available volumes implies an increasing need for labour and machinery. While these have so far not been considered as constraints in the availability assessments, mobilising more wood is likely to increase the need for skilled labour. To some extent, this lack could be alleviated by mechanisation and technological development. Technological development could also increase the potentials, if formerly technically unavailable or too-costly resources became available. For example, developing models to estimate the right time to harvest a site on sensitive soil could remove a technical constraint (Salmivaara et al. 2020).

Mobilising such potentially available volumes would mean a more intensive use of the EU's forest resources compared to the current situation. At the same time, the EU is trying to maintain and strengthen its forest carbon sinks. The EU's Land Use, Land-Use Change and Forestry Regulation (EU 2018) requires that the carbon sinks of a member state be compared to a reference level, assuming a continuation of historical (2000–2009) forest management practices. Should the future forest carbon sink be lower than the reference level, a member state would generate carbon debits. It will be important to assess how increasing wood use will affect the carbon balances of forests, wood products and through substitution effects.

Finally, even if ecological constraints are generally considered in biomass availability studies, higher harvesting levels could still affect carbon storage, biodiversity and other forest functions other than wood production. In order to protect biodiversity, the EU Biodiversity Strategy proposes to increase the area of protected forest (European Commission 2020). The strategy also aims to increase the quantity, quality and resilience of EU forests. It proposes to achieve this by planting 3 billion additional trees in the EU by 2030 and by establishing protected areas for at least 30% of the land in Europe, with stricter protection of European forests. The effect of this strategy on forest biomass availability remains to be seen.

5.4   Research Implications

There is a substantial body of literature relating to the harvesting potential of forest biomass in Europe. Despite the improved understanding this provides, important challenges remain relating to the availability of data, ownership structures and behaviour, and climate change (Barreiro et al. 2017; Nabuurs et al. 2019), all of which require further research.

Existing studies on biomass availability typically rely on national forest inventory data, with these data forming a solid basis for availability estimations (Vidal et al. 2016). In recent years, countries have shifted towards statistical inventories, which has improved the reliability and accuracy of the inventory results. However, forest inventories mostly rely on national definitions, which reduce their comparability, although progress is being made in overcoming this (Alberdi et al. 2016, 2020; Gschwantner et al. 2019). A key challenge relates to the availability of such inventory data, as they are not always readily available (Nabuurs et al. 2019). Improved availability would support 'top-down' assessments of biomass availability, which, together with 'bottom-up' assessments, provide important insights into biomass availability (Barreiro et al. 2017) and European forest resources more generally. An improved availability of data would also facilitate the increased use of remote-sensing-based data to provide up-to-date and large-scale information on Europe's forest resources (e.g. Moreno et al. 2017), and thus also biomass availability assessments.

It is evident that climate change will affect forests and forest biomass availability. Some European regions may benefit from the increased growth, while others will face reduced productivity or suffer from extreme events and natural disturbances (Lindner et al. 2014; Reyer et al. 2017; see also Chap. 3) and, thereby, the availability of wood. The frequency and intensity of forest disturbances are also likely to increase in the future (Seidl et al. 2014, 2017). Forest disturbances can cause strong peaks in biomass availability and will increase logistics costs, with the capacity of nearby industries potentially not being able to digest sudden supply peaks. For the industry, a constant supply of uniform quality is desirable, whilst for forest owners, forest damage means lowered timber quality and prices. Future climate change impacts (including disturbances) need to be included in long-term forest planning (Senf and Seidl 2020), as well as in studies assessing biomass availability.

To anticipate the impacts of climate change, strategies are being explored to improve the resilience of forests in the context of climate change. A key strategy is to increase species diversity—especially by increasing the share of broadleaved species—in temperate and boreal forest stands to improve forest resilience (Jactel et al. 2017; Astrup et al. 2018). Increasing species diversity will eventually affect the type of biomass assortments that will be available to the industry from forests. Further research is needed on how changes in biomass availability and quality may affect forest industries.

Forest owners typically have multiple objectives when managing their forests, and their attitude to mobilising more wood is unclear. While harvest probability generally increases with higher productivity of the region and species, there are important differences in harvesting decisions relating to local conditions, such as site accessibility, the state of the forest resource (age), specific subsidies, and the importance of other forest services (Schelhaas et al. 2018). Forest owner behaviour, and heterogeneity therein, should be considered in future studies on biomass availability (Blennow et al. 2014; Rinaldi et al. 2015; Stjepan et al. 2015; Sotirov et al. 2019).

5.5 Key Messages

- Existing studies indicate that a higher harvest level from EU forests could be sustained, but this will be associated with lower carbon storage in forest ecosystems, as well as impacts on other functions that forests have, including biodiversity.
- There are large differences between the European regions regarding harvesting potential and actual utilisation rate.
- The impacts of climate change on productivity are expected to vary across Europe. Climate change is expected to increase the frequency and intensity of forest disturbances, which can cause strong peaks in biomass availability and disrupt timber markets.

Box 5.1 Some Trends in the Global and the EU Forests

Antti Asikainen
Natural Resources Institute Finland, Joensuu, Finland

Forests cover 4,06 billion hectares (31%) of the world's total land area (FAO 2020b). Since 1990 the world has lost 178 million ha of forest, but the rate of net forest loss has been decreasing in recent decades. In Europe, Oceania and Asia, the forest area has been increasing whereas in Africa and South America it has been decreasing (FAO 2020b). Global drivers effecting the entire land-use sector are behind the changes of forest area. Increasing global population and changes in the diet have created growing demand for food production and land for farming and crazing. Commercial agriculture is the most important driver of deforestation followed by local agriculture, urban expansion, infrastructure and mining (Hosonuma et al. 2012). Forest degradation, however, is driven by timber harvesting opening the forest areas for low intensity farming and grazing and subsequent human induced fires (Hosonuma et al. 2012).

(continued)

Box 5.1 (continued)

In the EU, area of forest land has steadily increased since 1990's (Eurostat 2020). Hosonuma et al. classify developing countries in four forest transition (FT) phases. Pre transition countries have high forest cover and low deforestation rate, deforestation is at its highest in early and late transition phases, whereas in post transition phase forest cover starts to increase due to reforestation. Although this classification was designed for developing countries, it fits well to industrialized countries, too. EU countries in the last 50–100 years can be classified to post transition phase, where large reforestation programmes resulted to increasing forest cover.

The conversion of forest into other uses, mainly to farmland, pastures and cities, roads and other human infrastructures has been the major factor behind the negative climate impacts in land use sector. Thus, a central element for *climate smart forestry* is that forests stays as forests. Landowners and governments seek economic returns for their assets including land. When forest-based livelihoods offer less income, other land uses e.g. agriculture, mining and urban expansion take land. There are also reverse processes going on. For instance, in Finland in 1969 to 2002s due to overproduction of agricultural goods low value farm slots have been reforested. In total 240,000 hectares were replanted with the governmental support. More recently (2007–2013), the EU's rural development policy induced an increase of 1 to 2% of the forest area in some Member States/Regions such as ES – Asturias, ES - Castilla y León, ES – Galicia, HU, LT, UK – England, UK - Northern Ireland, and of 3% in UK – Scotland totaling c.a. 290,000 ha and even more is expected to be afforested by 2020 (Anon. 2017). Uruguay is an illustrative example of large-scale, market driven reforestation for the needs of rapidly growing pulp industry. Its forest area has increased 1990–2016 from 8000 km^2 to almost 19,000 km^2 (Anon. 2020).

Forest area as a proportion of total land area is a global indicator of the UN Sustainable Development Goals (SDGs). It is also included in the set of the EU SDG indicators used to monitor progress towards the SDGs in the EU context. It is important to recognize the effects of global trade into forest area. The EU has paid attention into its impacts on global land use induced by e.g. importing of soybean, wheat and other cereals and agricultural products (COM 2019).

Box 5.2 Huge Russian Forest Resources – A Reality or an Illusion?

Antti Mutanen and Sari Karvinen

Natural Resources Institute Finland, Joensuu, Finland

Based on a quick glance at the forest statistics, the forest resources of the Russian Federation (Russia) can be considered simply huge. According to the Food and Agriculture Organization's (FAO) most recent Global Forest Resource Assessment, Russia's forested area of 815 million ha is the largest of all the countries in the world, whilst the growing stock of 81.1 billion m^3 is the second largest after Brazil (FAO 2020c). Forests available for wood supply (FAWS, as defined by the FAO) account for over 80% of the total forest area and growing stock.

The easily created image of immense taiga providing virtually endless forest resources is, however, somewhat deceiving. A more detailed investigation of the statistics has revealed that the growth rate of Russian forests is low. The estimated net annual increment is about 1.0 billion m^3 in the total forest area and 850 million m^3 in the FAWS area, which translates into an average net increment rate of merely 1.3 $m^3ha^{-1}a^{-1}$ (Forest Europe 2020b; Roslesinforg 2021). For comparison, the average net increment rate (FAWS) is about 4.9 $m^3ha^{-1}a^{-1}$ in Finland and 4.8 $m^3ha^{-1}a^{-1}$ in Sweden (Forest Europe 2020b). The low growth rate of Russian forests is attributable to harsh climatic conditions (more than half of the forests are situated on permafrost soils), unfavourable age structures (about half of the coniferous forests are classed as mature and over-mature), as well as the prevailing forest management practices (extensive forestry based on large-scale clear-fellings and natural regeneration combined with a low level of tending seedling stands and intermediate fellings).

In addition to the growth rate, the utilisation rate of forests is also low in Russia. In the peak year of 2018, wood harvesting reached 240 million m^3, and the ratio of fellings to growth (FAWS) was 28% (Forest Europe. 2020b; FAOSTAT 2021). For comparison, the corresponding ratio was 71% in Finland and 79% in Sweden. The low ratio of fellings to growth is obviously far from being a desirable state of affairs in Russia. The administratively set annual allowable cut (AAC, raschetnaya lesoseka) for the whole country is currently 730 million m^3, or about 85% of the net annual increment (FAWS). The determination of the AAC is based on the characteristics of the forest, such as age and tree species composition. The AAC represents the level of wood harvesting that is sustainable, in terms of timber production and preserving the biodiversity and protective functions of the forests, which, according to the Russian forest-use classification system, belong to the exploitable and protective classes (Order of the Federal…). Moreover, the AAC can be considered the target level of wood harvesting in the state-owned forests, used, for example, as the basis for lease payments in leased forest areas (Forest Code…).

(continued)

Box 5.2 (continued)

The current level of the AAC has been criticised by Russian experts as being an unrealistic overestimate of the wood production potential (Shvarts 2018; Strategiya razvitiya lesnogo…). The foremost reason for this is the inadequate infrastructure, especially the lack of a comprehensive forest road network and missing railway connections, which means that vast expanses of Russia's forests are currently simply unreachable. In addition, the AAC does not take into account the natural conditions, such as slopes, forest quality or use restrictions, adequately. It has been estimated that, without a considerable investment in infrastructure, the realistic AAC for the whole of Russia is currently about 340 million m^3 (Strategiya razvitiya lesnogo…).

In Fig. Box 5.1, the area covered by forests is contrasted with the forest area considered accessible for transport. The accessible forests are concentrated on the European part of Russia, where the infrastructure is relatively well developed and where the majority of the production capacity of the Russian forest industry is located. However, the most easily accessible forests, within a reasonable transportation distance to forest industry complexes, were felled decades ago, which, combined with poor forest management practices, has led to a deterioration in forest quality; that is, coniferous forests have been replaced by deciduous ones. Moreover, in some regions, a marked share of the coniferous forests is accessible only during winter due to the low bearing capacity of the forest soils, which, under warming climate conditions, makes wood procurement vulnerable (Goltsev and Lopatin 2013). Thus, in many regions in the European part of Russia, the forest industry is suffering from an inadequate supply of coniferous timber assortments, especially sawlogs, and simultaneously, there is practically no demand for deciduous pulpwood (State Council…2013). This situation has led to the overexploitation of the remaining coniferous forests, while natural losses have grown rapidly in the deciduous forests.

Various stakeholders in the Russian forest sector, including wood processors, logging companies, the Federal Forestry Agency (Rosleskhoz), regional forest management bodies, and environmental non-governmental organisations, consider intensive sustainable forest management (ISFM) to be a viable means for tackling the problems relating to the low productivity and deteriorating quality of the forests (Gosudarstvennaya programma Rossiyskoy…; Shmatkov 2013a, b; Intensivnoye lesnoye khozyaystvo…2015). The forest management practices under the ISFM are basically the same as those used in the Nordic countries—mainly artificial regeneration, active tending of seedling stands, first thinnings and other intermediate fellings, followed by the final felling. By applying ISFM, more wood could be produced per hectare and per year than currently, which would help to secure the wood supply for the forest industry, while simultaneously easing the pressure on opening intact forest areas to logging.

(continued)

Box 5.2 (continued)

Fig. Box 5.1 Forests and potentially productive forests accessible for transport. The area of forested land (680 million ha) is based on Landsat data and includes areas with forest cover (canopy cover) greater than 20% and potential regeneration areas, such as logging sites and burnt areas. The FAO's definition of a forest sets the canopy cover threshold at 10%, hence the difference in forest area estimates (680 vs. 815 million ha). Potentially productive forests accessible for transport are forest areas (excluding nature-protection areas and intact forest landscapes) in which the long-term potential average increment is more than 1 m³ha⁻¹ (based on the Moderate Resolution Imaging Spectroradiometer (MODIS) Net Primary Production product) and the forest transportation distance to an existing road network is less than 1 km (Lopatin 2017)

(continued)

Box 5.2 (continued)

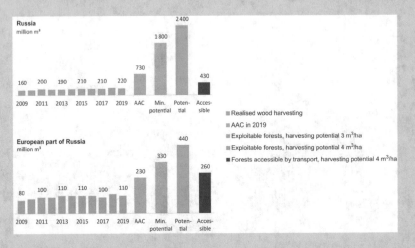

Fig. Box 5.2 Realised wood harvesting, the 2019 AAC, harvesting potentials under the ISFM in exploitable forests and accessible forests in the whole of Russia and in the European part of Russia. Accessible forests are the same as in Fig. 5.1. (Sources: Saint-Petersburg Forestry...2015; Lopatin 2017; Finnish Forest Statistics 2020; Rosleskhoz 2020; Roslesinforg 2021; Yedinaya mezhvedomstvennaya 2021)

Russian experts have estimated that the ISFM could raise the harvesting volumes to 3–4 m^3ha^{-1} in the Northern Taiga Zone and to 5–6 m^3ha^{-1} in the Central and Southern Taiga Zones (Saint-Petersburg Forestry...2015). For comparison, the harvesting intensity was 4.0 m^3ha^{-1} (FAWS) in Finland in 2019 (Finnish Forest Statistics 2020). Figure Box 5.2. demonstrates the hypothetical harvesting volumes for different harvesting intensities and for different forest areas under the ISFM. If the ISFM was applied in all the forests classed as exploitable, according to the Russian forest-use classification system, and assuming that 4 m^3ha^{-1} could be harvested, the total harvested volume would be 2.4 billion m^3 across the whole of Russia; that is, almost tenfold the current harvesting volumes or more than threefold the AAC. If the ISFM was applied to the forest area considered accessible with the existing infrastructure, the harvesting level would be 430 million m^3, or roughly 60% of the AAC and 1.9 times higher than the current harvesting volumes. On the European part of Russia, applying the ISFM to the accessible forests would hypothetically lead to a harvesting volume of twice the current fellings and equal to the AAC. Thus, changing the forestry doctrine from extensive to intensive could greatly increase forest productivity in Russia, as well as the sustainable harvesting volumes. In several pilot areas, such as the Republics

(continued)

Box 5.2 (continued)

of Komi and Karelia, Arkhangelsk, Vologda, Irkutsk, Kirov and the Leningrad region, forestry norms already allow ISFM practices. However, there are no statistics on how large a scale the leaseholders have adopted, or are planning to adopt, the ISFM methods.

Many factors need to be considered when interpreting the harvesting potentials presented in Fig. 5.2. Although a forest area may be classed as accessible, there may not be any demand for the wood due to the tree species being wrong or the transportation distance to a processing plant being too long. In other words, besides the physical accessibility, economic accessibility is also needed. Moreover, a shift to the forest management practices of the ISFM requires investments in forest regeneration, the tending of seedling stands and a forest road network, a skilled workforce, and time. Taking the example of the road network, a 'conditionally adequate' road density for forestry is, on average, 5 m ha^{-1}, according to a Russian assessment (State Council...2013). The current forest road density is about 3 m ha^{-1}, thus the construction of 1 million km of new roads would be required to reach an adequate density in Russia's exploitable forests. The costs would be EUR 13 billion, using the Finnish pricing level (Finnish Forest Statistics 2020). In practice, the costs would be more substantial, as road construction costs are higher in Russia (Petrunin 2013; Havimo et al. 2017). To achieve the forest road density in Finland (10 m ha^{-1}; Uotila and Viitala 2000), more than 4 million km of new roads and EUR 57 billion would be required. For comparison, in 2019, forestry financing totalled EUR 1.1 billion, of which EUR 0.3 billion was invested by the forest leaseholders (Accounts Chamber...2020).

How to create incentives for forest leaseholders to invest in management activities with payback times that will, at best, be decades in the future, when there are also no guarantees that the lease period will be continued? Where to find enough workers to execute labour-intensive operations, such as tending seedling stands and thinnings, which are not that common in Russia? At present, using Finland as an example, the number of forestry workers in relation to the exploitable forest area is four times higher than in Russia, and even though there would be a willingness and the assets available to invest in forest management as well as the necessary workforce, materialisation of the full harvesting potential provided by the ISFM would take decades.

Russian forests are vast by area and volume, and for decades, the realised harvesting volumes have been far less than the AAC, the growth or the myriad different kinds of harvesting potential estimates. It would be tempting to interpret that there is a substantial – even astronomical – potential to increase the material use of forests in Russia. However, it is unrealistic to assume that the harvesting levels could be raised considerably in the short or medium terms.

(continued)

Box 5.2 (continued)

This fact has also been acknowledged in the Russian forest sector development strategy, whose most positive scenario for the harvesting level of 2030 is 286 million m³. In the long term, it would be quite possible to reinforce both the growth and harvesting volumes greatly. However, in order to realise this development path, several challenges have to be overcome. Infrastructure has to be developed, forest management rethought, the use of wood diversified, the operating environment stabilised, etc. Naturally, money is needed, but also politically sensitive issues, such as the private ownership of forests and the role of foreign investors, have to be discussed at a profound level.

The vast forest resources have always tempted non-Russian wood processors to source feedstock from Russia. However, the task has never been an easy one. During the last few decades, Russia has aimed at restricting roundwood exports, and the means to do this have included protective export duties, quotas and different regulations that complicate export procedures (Karvinen et al. 2019). In fact, after a years-long period of relatively stable trade conditions, roundwood sourcing from Russia to Europe will again become significantly more difficult in the very near future. A plan to ban exports of softwood and valuable hardwood from Russia was announced at the end of 2020 (Presidential instructions…2020). As a consequence, tariff quotas for softwood (spruce and pine) are to be removed from 2022 onwards (Government Decree…396). Under the valid regulation on customs duties, this will lead to prohibitive duties – for softwood, a minimum of €55 m $^{-3}$ (Government Decree…754). A new development that has been raised is the possibility of restricting the export of softwood chips. It has been suggested that softwood chips be added to the products crucial for the internal Russian market on which temporary export restrictions or an export ban can be imposed (Draft of Government Decree…). The message is quite clear – harvested roundwood should be processed in Russia to create value added.

References

Alberdi I, Michalak R, Fischer C, Gasparini P, Brändli U-B, Tomter SM et al (2016) Towards harmonized assessment of European forest availability for wood supply in Europe. For Policy Econ 70:20–29

Alberdi I, Bender S, Riedel T, Avitable V, Boriaud O, Bosela M et al (2020) Assessing forest availability for wood supply in Europe. For Policy Econ 111:102032. https://doi.org/10.1016/j.forpol.2019.102032

Anttila P, Nivala V, Salminen O, Hurskainen M, Kärki J, Lindroos TJ, Asikainen A (2018) Regional balance of forest chip supply and demand in Finland in 2030. Silva Fenn 52(2):10.14214/sf.9902

Astrup R, Bernier PY, Genet H, Lutz DA, Bright RM (2018) A sensible climate solution for the boreal forest. Nat Clim Chang 8:11–12

Barreiro S, Schelhaas M-J, McRoberts RE, Kändler G (eds) (2017) Forest inventory-based projection systems for wood and biomass availability. Springer, Cham

Blennow K, Persson E, Lindner M, Faias SP, Hanewinkel M (2014) Forest owner motivations and attitudes towards supplying biomass for energy in Europe. Biomass Bioenergy 67:223–230

Brosowski A, Thrän D, Mantau U, Mahro B, Erdmann G, Adler P et al (2016) A review of biomass potential and current utilisation – status quo for 93 biogenic wastes and residues in Germany. Biomass Bioenergy 95:257–272

Camia A, Robert N, Jonsson R, Pilli R, García-Condado S, López-Lozano R, … Giuntoli J (2018) Biomass production, supply, uses and flows in the European Union: First results from an integrated assessment, EUR 28993 EN. Publications Office of the European Union, Luxembourg. 10.2760/539520

Cazzaniga NE, Jonsson R, Palermo D, Camia A (2019a) Sankey diagrams of woody biomass flows in the EU-28. EC Joint Research Centre, Publications Office of the European Union, Luxembourg. https://doi.org/10.2760/227292

Cazzaniga NE, Jonsson R, Pilli R, Camia A (2019b) Wood resource balances of EU-28 and member states. EC Joint Research Centre, Publications Office of the European Union, Luxembourg https://doi.org/10.2760/020267

de Vries W, Solberg S, Dobbertin M, Sterba H, Laubhann D, van Oijen M et al (2009) The impact of nitrogen deposition on carbon sequestration by European forests and heathlands. For Ecol Manag 258:1814–1823

Dees M, Elbersen B, Fitzgerald J, Vis M, Anttila P, Forsell N, … Höhl M (2017) Atlas with regional cost supply biomass potentials for EU 28, Western Balkan countries, Moldavia, Turkey and Ukraine.   https://www.s2biom.eu/images/Publications/D1.8_S2Biom_Atlas_of_regional_cost_supply_biomass_potential_Final.pdf. Accessed 21 May 2021

Di Fulvio F, Forsell N, Lindroos O, Korosuo A, Gusti M (2016) Spatially explicit assessment of roundwood and logging residues availability and costs for the EU28. Scand J Forest Res 31:691–707

Energimyndigheten (2020) Production of unprocessed forest fuels by fuel type and feedstock origin.   https://pxexternal.energimyndigheten.se/pxweb/en/?rxid=2c91707b-7c5e-40 5bb132-3aac75a4a172. Accessed 10 June 2020

Etzold S, Ferretti M, Reinds GJ, Solberg S, Gessler A, Waldner P et al (2020) Nitrogen deposition is the most important environmental driver of growth of pure, even-aged and managed European forests. For Ecol Manag. https://doi.org/10.1016/j.foreco.2019.117762

EU (2018) Regulation (EU) 2018/841 of the European Parliament and of the Council of 30 May 2018 on the inclusion of greenhouse gas emissions and removals from land use, land-use change and forestry in the 2030 climate and energy framework, and amending Regulation (EU) No 525/2013 and Decision No 529/2013/EU. https://eur-lex.europa.eu/eli/reg/2018/841. Accessed 21 May 2021

European Commission (2020) EU biodiversity strategy for 2030. https://eur-lex.europa.eu/legal-content/EN/TXT/?uri=CELEX:52020DC0380

FAO (2020) FAOSTAT. Forestry production and trade. http://www.fao.org/faostat/en/#data/FO. Accessed 8 June 2020

FAO, ITTO and United Nations (2020) Forest product conversion factors. Rome. https://doi.org/10.4060/ca7952en

Flechard CR, Ibrom A, Skiba UM, de Vries W, van Oijen M, Cameron DR et al (2020) Carbon–nitrogen interactions in European forests and semi-natural vegetation – part 1: fluxes and budgets of carbon, nitrogen and greenhouse gases from ecosystem monitoring and modelling. Biogeosciences 17:1583–1620

Forest Europe (2020a) State of Europe's Forests 2020. https://foresteurope.org/state-europes-forests-2020. Accessed 21 Dec 2020

Forzieri G, Pecchi M, Girardello M, Mauri A, Klaus M, Nikolov C et al (2020) A spatially explicit database of wind disturbances in European forests over the period 2000–2018. Earth Syst Sci Data 12:257–276

Gardiner B, Blennow K, Carnus J-M, Fleischer P, Ingemarson F, Landmann G, ... Usbeck T (2010) Destructive storms in European forests: Past and forthcoming impacts. Final report to the European Commission. DG Environment. http://ec.europa.eu/environment/forests/pdf/ STORMS%20Final_Report.pdf. Accessed 24 May 2021

Gold S, Korotkov A, Sasse V (2006) The development of European forest resources, 1950 to 2000. For Policy Econ 8:183–192

Gschwantner T, Alberdi I, Balázs A, Bauwens S, Bender S, Borota D et al (2019) Harmonisation of stem volume estimates in European National Forest Inventories. Ann For Sci. https://doi. org/10.1007/s13595-019-0800-8

Jactel H, Bauhus J, Boberg J, Bonal D, Castagneyrol B, Gardiner B et al (2017) Tree diversity drives Forest stand resistance to natural disturbances. Curr For Rep 3:223–243

Jochem D, Weimar H, Bösch M, Mantau U, Dieter M (2015) Estimation of wood removals and fellings in Germany: a calculation approach based on the amount of used roundwood. Eur J For Res 134:869–888

Jonsson R, Blujdea V, Fiorese G, Pilli R, Rinaldi F, Baranzelli C, Camia A (2018) Outlook of the European forest- based sector: Forest growth, harvest demand, wood-product markets, and forest carbon dynamics implications. iForest 11:315–328

Kärkkäinen L, Haakana H, Hirvelä H, Lempinen R, Packalen T (2020) Assessing the impacts of land-use zoning decisions on the supply of Forest ecosystem services. Forests. https://doi. org/10.3390/f11090931

Leskinen P, Lindner M, Verkerk PJ, Nabuurs GJ, Van Brusselen J, Kulikova E, Hassegawa M, Lerink B (eds) (2020) Russian forests and climate change. What science can tell us 11. European Forest Institute. https://doi.org/10.36333/wsctu11

Lindner M, Dees M, Anttila P, Verkerk P, Fitzgerald J, Datta P, Glavonjic B, Prinz R, Zudin S (2017) Assessing Lignocellulosic Biomass Potentials From Forests and Industry. In: Panotsou C (Ed.), Modeling and Optimization of Biomass Supply Chains: Top-Down and Bottom-up Assessment for Agricultural (pp. 127–167). Forest and Waste Feedstock. Academic Press

Moreno A, Neumann M, Hasenauer H (2017) Forest structures across Europe. Geosci Data J 4:17–28

Nabuurs GJ, Verweij P, van Eupen M, Pérez-Soba M, Pülzi H, Hendriks K (2019) Next-generation information to support a sustainable course for European forests. Nat Sustain 2:815–818

Natural Resources Institute Finland (2020) Solid wood fuel consumption in heating and power plants by region. Statistical database. http://statdb.luke.fi/PXWeb/pxweb/en/LUKE/ LUKE04%20Metsa04%20Talous10%20Puun%20energiakaytto/01a_Laitos_ekaytto_maak. px/?rxid=9a0b5502-10d0-4f84-8ac5-aef44ea17fda Accessed 10 June 2020

Pretzsch H, Biber P, Schütze G, Uhl E, Rötzer T (2014) Forest stand growth dynamics in Central Europe have accelerated since 1870. Nat Commun. https://doi.org/10.1038/ncomms5967

Reyer C, Bathgate S, Blennow K, Borges JG, Bugmann H, Delzon S et al (2017) Are forest disturbances amplifying or canceling out climate change-induced productivity changes in European forests? Environ Res Lett. https://doi.org/10.1088/1748-9326/aa5ef1

Rinaldi F, Jonsson R, Sallnäs O, Trubins R (2015) Behavioral modelling in a decision support system. Forests 6:311–327

Salmivaara A, Launiainen S, Perttunen J, Nevalainen P, Pohjankukka J, Ala-Ilomáki J et al (2020) Towards dynamic forest trafficability prediction using open spatial data, hydrological modelling and sensor technology. Forestry 93:662–674

Schelhaas MJ, Fridman J, Hengeveld GM, Henttonen HM, Lehtonen A, Kies U et al (2018) Actual European forest management by region, tree species and owner based on 714,000 re-measured trees in national forest inventories. PLoS One. https://doi.org/10.1371/journal.pone.0207151

Seidl R, Schelhaas MJ, Rammer W, Verkerk PJ (2014) Increasing forest disturbances in Europe and their impact on carbon storage. Nat Clim Chang 4:806–810

Seidl R, Thom D, Kautz M, Martin-Benito D, Peltoniemi M, Vacchiano G et al (2017) Forest disturbances under climate change. Nat Clim Chang 7:395–402

Senf C, Seidl R (2020) Mapping the forest disturbance regimes of Europe. Nat Sustain. https://doi. org/10.1038/s41893-020-00609-y

Sotirov M, Sallnäs O, Eriksson LO (2019) Forest owner behavioral models, policy changes, and forest management. An agent-based framework for studying the provision of forest ecosystem goods and services at the landscape level. For Policy Econ 103:79–89

Stjepan P, Mersudin A, Dženan B, Petrovic N, Stojanovska M, Marceta D, Malovrh SP (2015) Private forest owners' willingness to supply woody biomass in selected south-eastern European countries. Biomass Bioenergy 81:144–153

Verkerk PJ, Anttila P, Eggers J, Lindner M, Asikainen A (2011) The realisable potential supply of woody biomass from forests in the European Union. For Ecol Manag 261:2007–2015

Verkerk PJ, Fitzgerald JB, Datta P, Dees M, Hengeveld GM, Lindner M, Zudin S (2019) Spatial distribution of the potential forest biomass availability in Europe. For Ecosyst. https://doi.org/10.1186/s40663-019-0163-5

Vidal C, Alberdi IA, Hernández Mateo L, Redmond JJ (2016) National Forest Inventories: assessment of wood availability and use. Springer, Cham

Vilén T, Cienciala E, Schelhaas MJ, Verkerk H, Lindner M, Peltola H (2016) Increasing carbon sinks in European forests: effects of afforestation and changes in mean growing stock volume. Forestry 89:82–90

Reference from Box 5.1

Anon (2017) Evaluation study of the forestry measures under rural development technical report. September 2017 https://doi.org/10.2762/06029. 176 p

Anon (2020) World data atlas. https://knoema.com/atlas/Uruguay/topics/Land-Use/Area/Forest-area

FAO (2020b) Global forest resource assessment 2020. Key findings. 14 .COM 2019. Stepping up EU Action to Protect and Restore the World's Forests. COM 2019 352 final. 21 p

Hosonuma N, Herold M, De Sy V, De Fries RS, Brockman M, Verchot L, Angelsen A, Romijn E (2012) An assessment of deforestation and forest degradation drivers in developing countries. Environ Res Lett 7(2012):12. https://doi.org/10.1088/1748-9326/7/4/044009

Reference from Box 5.2

Accounts Chamber of the Russian Federation (2020). Otchet o rezul'tatakh sovmestnogo ekpertno- analiticheskogo meropriyatiya "Analiz effektivnosti ispol'zovaniya lesnykh resursov Rossiyskoy Federatsii v 2016–2018 godakh" [Report on the results of a common expert-analytical measure "The analysis of efficiency of using forest resources in the Russian Federation in 2016–2018"]

FAO (2020c) Global Forest Resources Assessment 2020: Main report. Rome. https://doi.org/10.4060/ca9825en

FAOSTAT (2021) Forest product statistics. Forestry production and trade. http://www.fao.org/faostat/en/#data/FO. Accessed 17 May 2021

Finnish Forest Statistics (2020) Natural Resources Institute Finland

Forest code of the Russian Federation 04.12.2006 N 200-FZ

Forest Europe (2020b) State of Europe's Forests 2020. Ministerial Conference on the Protection of Forests in Europe

Goltsev V, Lopatin E (2013) The impact of climate change on the technical accessibility of forests in the Tikhvin District of the Leningrad region of Russia. Int J For Eng 24(2):148–160. https://doi.org/10.1080/19132220.2013.792150

Gosudarstvennaya programma Rossiyskoy Federatsii "Razvitiye lesnogo khozyaystva" na 2013–2020 gody [State programme of the Russian Federation "Development of forestry" 2013–2020] Government Decree of the Russian Federation 18.03.2021 No. 396

Government Decree of the Russian Federation 30.8.2013 No. 754 (including changes upto 18.3.2021)

Havimo M, Mönkönen P, Lopatin E, Dahlin B (2017) Optimising forest road planning to maximise the mobilisation of wood biomass resources in Northwest Russia. Biofuels 8(4):501–514. https://doi.org/10.1080/17597269.2017.1302664

Presidential instructions following the meeting on the development and decriminalization of the forest sector. 6.11.2020. http://www.kremlin.ru/acts/assignments/orders/64379. Accessed 17 May 2021

Draft of Government Decree "Changing the list of products especially important for the internal market of the Russian Federation…". https://regulation.gov.ru/p/112935. Accessed 17 May 2021

Intensivnoye lesnoye khozyaystvo: Obyazannost ili osoznannaya neobkhodimost? [Intensive forestry: Responsibility or the recognition of necessity?] 2015. Ustoychivoye lesopol'zovaniye 1(41):34–41. https://wwf.ru/upload/iblock/b01/07-_5_.pdf. Accessed 17 May 2021

Karvinen S, Mutanen A, Petrov V (2019) Effects of the export restrictions on birch log market in Northwest Russia. Balt For 25(1):105–112. https://doi.org/10.46490/vol25iss1pp105

Lopatin E (2017) Ranzhirovaniye uchastkov lesov Rossii po vozmozhnosti vnedreniya metodov intensivnogo ustoichivogo lesnogo khozyaystva [Ranking of Russian forest areas possible for intensive sustainable forest management]. Ustoichivoye lesopol'zovaniye No. 4(52):2–7. https://wwf.ru/upload/iblock/b3f/01.pdf. Accessed 17 May 2021

Order of the Federal Forestry Agency 27.5.2011 N 191

Petrunin N (2013) Lesnoye bezdorozh'e (Roadless forest). "Derewo.ru" 5:30–33

Roslesinforg (2021) Lesa Rossii [Russian forests]. https://roslesinforg.ru/atlas. Accessed 17 May 2021

Rosleskhoz (2020) Svedeniya o zemlyakh lesnogo fonda [Information on the forest fund land]. http://rosleshoz.gov.ru/opendata/7705598840-ForestFund. Accessed 15 April 2020

Saint-Petersburg Forestry Research Institute (2015) Kontseptsiya intensivnogo ispol'zovaniya i vosproizvodstva lesov [The concept of intensive use and regeneration of forests]. Saint Petersburg Forestry Research Institute. FBU "SPBNIILH", St. Petersburg

Shmatkov N (ed) (2013a) Intensivnoye ustoichivoye lesnoye khozyaystvo: Bar'ery i perspektivy razvitiya [Intensive sustainable forestry: Barriers and development perspectives]. WWF Russia, Moscow

Shmatkov N (ed) (2013b). Strategiya razrabotki sistemy lesokhozyaystvennykh i prirodookhrannykh normativov dlya Sredne-tayezhnogo lesnogo rayona s tselyu vnedreniya modeli ustoichivogo intensivnogo lesnogo khozyaystva [Strategy for compiling the forestry and nature protection norms for the central taiga forest zone with the aim of implementing sustainable and intensive forestry]. Materials of the round table 18 March 2013, St. Petersburg. WWF, Moscow

Shvarts E, Shmatkov H, Kobyakov K, Rodionov A, Yaroshenko A (2018) Nekotorye prichiny krizisa lesnogo sektora i puti vykhoda iz nego [Some reasons for the crisis in the forest sector and ways out]. Ustoichivoye lesopol'zovaniye No. 3(55):4–16. https://wwf.ru/upload/iblock/e0b/02.pdf. Accessed 15 Apr 2020

State Council of the Russian Federation (2013). Doklad o povyshenii effektivnosti lesnogo kompleksa [Report on increasing efficiency of the forest sector]. State Council of the Russian Federation

Strategiya razvitiya lesnogo kompleksa Rossiyskoy Rossiyskoy Federatsii na period do 2030 goda [Forest sector development strategy of the Russian Federation until 2030]. Confirmed by Government Decree 11.2.2021 No. 312

Uotila E, Viitala E-J (2000) Tietiheys metsätalouden maalla. Metsätieteen aikakauskirja 1/2000: 19–33 (in Finnish). https://metsatieteenaikakauskirja.fi/pdf/article6910.pdf. Accessed 15 April 2020

Transliteration: BGN/PCGN romanisation of Russian

Yedinaya mezhvedomstvennaya informatsionno-statisticheskaya Sistema (2021) [The Unified Interdepartmental Statistical Information System]. https://fedstat.ru. Accessed 17 May 2021

Chapter 6
Carbon Sequestration and Storage in European Forests

Antti Kilpeläinen and Heli Peltola

Abstract European forests have been acting as a significant carbon sink for the last few decades. However, there are significant distinctions among the forest carbon sinks in different parts of Europe due to differences in the area and structure of the forests, and the harvesting intensity of these. In many European countries, the forest area has increased through natural forest expansion and the afforestation of low-productivity agricultural lands. Changing environmental conditions and improved forest management practices have also increased the carbon sequestration and storage in forests in different regions. The future development of carbon sequestration and storage in European forests will be affected both by the intensity of forest management and harvesting (related to future wood demand) and the severity of climate change and the associated increase in natural forest disturbances. Climate change may also affect the carbon dynamics of forests in different ways, depending on geographical region. Therefore, many uncertainties exist in the future development of carbon sequestration and storage in European forests, and their contribution to climate change mitigation. The demand for multiple ecosystem services, and differences in national and international strategies and policies (e.g. the European Green Deal, climate and biodiversity policies), may also affect the future development of carbon sinks in European forests.

Keywords Carbon balance · Carbon sink · Carbon source · Carbon stock · Climate change · Forest ecosystem · Forest management · Growth · Net ecosystem CO_2 exchange · Mitigation

A. Kilpeläinen (✉) · H. Peltola
University of Eastern Finland, Joensuu, Finland
e-mail: antti.kilpelainen@uef.fi

© The Author(s) 2022
L. Hetemäki et al. (eds.), *Forest Bioeconomy and Climate Change*, Managing
Forest Ecosystems 42, https://doi.org/10.1007/978-3-030-99206-4_6

6.1   Current Carbon Storage and Sink

Forests can contribute significantly to the global carbon cycle and climate change mitigation by sequestering carbon from the atmosphere and storing it in forests (forest biomass and soil) and in wood-based products (with long life-cycles), and also through the use of forest biomass to substitute for fossil-fuel-intensive materials, products and fossil energy (Nabuurs et al. 2017; Leskinen et al. 2018). This is also the case in Europe, where the majority of forests are managed. Forest management has largely influenced the present tree species composition (Spiecker 2003) and wood production potential (Rytter et al. 2016; Verkerk et al. 2019) of forests, and will continue to do so for the coming decades (e.g. Koehl et al. 2010; Lindner et al. 2014).

In Europe, the forest area and carbon storage have both increased since the 1950s for several reasons. The forest area has increased by about 30% between 1950 and 2000, and by 9% since 1990 up to the present (Forest Europe 2020). This has occurred through natural forest expansion and the afforestation of low-productivity agricultural lands (e.g. Gold et al. 2006; Forest Europe 2015; Vilén et al. 2016). The ratio of annual harvested timber to the total annual increment of forests is below 80% across Europe, remaining relatively stable for most countries for the last few decades (European Environmental Agency [EEA] 2017). Additionally, improved forest management practices and changing environmental conditions (e.g. nitrogen deposition, climate warming and the elevation of atmospheric CO_2 concentrations) have increased the carbon sequestration and storage in European forests (e.g. Pretzsch et al. 2014; Etzold et al. 2020). However, the growing (carbon) stock of European forests has clearly increased more rapidly over the last few decades than the forest area (e.g. 17.5 million ha between 1990 and 2015), as the average volume per hectare has been increasing.

However, there are significant distinctions among the forest carbon sinks in different parts of Europe due to large differences in the forest area and structure (age and tree species composition). These are related to differences in the prevailing climatic and site conditions, the intensity of past and current forest management activities, and the level of socioeconomic development (EEA 2016). In Northern Europe, where the share of forest area is higher than in other parts of Europe, the forest landscapes are dominated by mainly coniferous, (very often) single-species and even-aged forests. In Central and Southern Europe, broadleaved deciduous and mixed evergreen forests are more common (Forest Europe 2020). Overall, the forests are more productive and have higher volumes of growing stock in Central Europe than in other parts of Europe. Forest productivity is, nowadays, limited by the length of the growing season and the relatively low summer temperatures in Northern Europe, whereas in Southern Europe, it is limited by water availability, with many forests also being located on sites with low potential for wood production.

The prevailing environmental conditions, current forest structure, management traditions and different socioeconomic factors have also affected the intensity of forest management. Management intensity varies from fully protective for biodiversity conservation, to uneven- and even-aged rotation forestry, which affects forest carbon sequestration and storage. Forest ownership structures, and targets set for forest management and its possible constraints, have also, together, affected the intensity of forest management and harvesting, affecting the development of carbon sinks and storage and the wood production potential of European forests (Rytter et al. 2016; Verkerk et al. 2019). Currently, ca. 50% of forests in the EU are privately owned, with about 16 million private forest owners (Nabuurs et al. 2015). In forest management, different ecosystem services may also be emphasised to a greater degree, depending on set targets and constraints in different regions (Hengeveld et al. 2012; EEA 2016; Forest Europe 2020).

The growing (carbon) stock of European forests is currently double what it was in the 1990s (Forest Europe 2020). The carbon-stock increases in forests and wood products, and the average annual sequestration of carbon in the forest biomass, was 155 million t in 2020 (Forest Europe 2020). Currently, EU forests sequester ca. 10% of Europe's greenhouse gas (GHG) emissions (Forest Europe 2020). When considering the carbon storage in wood products (an additional ca. 12 Tg C year^{-1}) and the substitution effects of the forest sector, ca. 3% of the total GHG emissions in the EU28 are avoided (Nabuurs et al. 2015). Furthermore, woody biomass provides ca. 6% of the energy consumed in the EU (Eurostat 2020). On the other hand, the first signs of saturation in the European forest carbon sink were recognised in the 2010s (Nabuurs et al. 2013). Despite this, the European forest carbon sink is still projected to last for decades. However, there may be a need to adapt forest management and utilisation strategies to promote the sequestration of carbon in forest sinks under the changing climatic conditions. Whether the carbon sink contained in European forests (and the broader forest sector) will remain at the same level as today, or increase/decrease in the future, will strongly depend on changes in the forest area and structure, the intensity of management and harvesting, and the severity of climate change and the associated increase in natural disturbances in different parts of Europe.

6.2 Dynamics of Carbon Sequestration and Storage in a Forest Ecosystem

6.2.1 Basic Concepts of Carbon Dynamics in a Forest Ecosystem

The carbon dynamics in a forest ecosystem comprise the carbon uptake by trees (and ground vegetation) in the above- and belowground forest biomass, and carbon release through the autotrophic (metabolism of organic matter by plants) and heterotrophic (metabolism of organic matter by bacteria, fungi and animals) respiration. The forest ecosystem is a carbon sink if it absorbs more carbon from the atmosphere than it emits, resulting in an increase in the carbon storage of the forest (forest biomass and soil). The carbon dynamics of a forest ecosystem are controlled

Table 6.1 Commonly used basic concepts of the sources, sinks and storage of carbon in a forest ecosystem

Carbon sequestration	Capture of CO_2 from the atmosphere and its transformation into biomass through photosynthesis
Carbon storage	Amount of carbon (stock) in the forest biomass and soil that has been removed from the atmosphere and stored in a forest ecosystem through carbon sequestration
Carbon sink	A forest ecosystem is a carbon sink if it absorbs more carbon from the atmosphere than it emits, resulting in an increase in carbon storage in the forest ecosystem. The net ecosystem exchange is negative (NEE = NPP – RH, <0)
Carbon source	A forest ecosystem is a carbon source if it emits more carbon into the atmosphere than it absorbs, resulting in the consequent reduction of carbon storage in the forest ecosystem. The net ecosystem exchange is positive (NEE = NPP – RH, >0)
Carbon balance	The carbon balance (NEE) of a forest ecosystem refers to the sum of carbon absorbed by and emitted from the forest ecosystem. If the carbon absorption is equal to the carbon emission, the carbon balance is zero

NEE net ecosystem CO_2 exchange, *NPP* net primary production, *RH* heterotrophic soil respiration

by environmental (climate, site) conditions, and the structure (age, stocking, tree species composition, etc.) and functioning of the forest ecosystem.

The carbon sequestration and stock of forest biomass may vary greatly in a forest ecosystem over time, these are controlled by the initial stand characteristics, the type and intensity of management (e.g. forest regeneration material, thinning and fertilisation) (Routa et al. 2019) and the length of the rotation period (Lundmark et al. 2018) or other time period being considered. The carbon stock in soil is generally relatively stable, although it is affected by carbon inputs from litter fall and carbon outputs from the decay of litter and humus, the latter representing earlier litter input of unrecognisable origin (Kellomäki et al. 2008). The decomposition of old humus and litter contributes significantly to soil carbon emissions at the beginning of the rotation period, but in the later stages of stand development, the decay of new litter contributes more (e.g. Kilpeläinen et al. 2011). Generally, for most of the duration of stand development, the stands act as carbon sinks (Table 6.1).

6.2.2 Management Effects on the Carbon Dynamics of a Forest Ecosystem

Management intensity affects the carbon sequestration and stocks in forest ecosystems through changing the structure and functioning of an ecosystem. A managed forest ecosystem sequesters carbon as trees grow, but loses carbon in harvesting. By comparison, in unmanaged forest ecosystems (e.g. old-growth forests), the carbon dynamics are affected by the age structure, the mortality of mature trees, natural regeneration and the ingrowth of seedlings in canopy gaps (Luyssaert et al. 2008). The annual growth rate of trees can be higher in managed than in unmanaged (intact) forest ecosystems, but the carbon sink is lower due to harvesting (Kellomäki 2017; Moomaw et al. 2020). Older forest stands can store more carbon, but the rate at which they remove additional carbon from the atmosphere is substantially lower, and can even become negative as the mortality increases and exceeds the regrowth (Gundersen et al. 2021). On the other hand, devastating abiotic (e.g. wind storms and forest fires) and biotic (e.g. insect outbreaks) disturbances may cause a sudden decrease in carbon sequestration and storage in forest ecosystems.

The use of appropriate, site-specific regeneration methods and materials (e.g. improved regeneration materials with better growth rates and survival), the proper timing and intensity of pre-commercial and commercial thinnings, and forest fertilisation on sites with limited nutrient availability, have been proposed as ways of increasing carbon sequestration (and timber production) over one rotation in boreal forests (e.g. Nilsen 2001; Saarsalmi and Mälkönen 2001; Bergh et al. 2014; Haapanen et al. 2015; Hynynen et al. 2015). According to Olsson et al. (2005), in addition to forest productivity, nitrogen fertilisation may also increase the sink and storage of carbon in upland (mineral) soils in Norway spruce stands due to the simultaneous increase in litter production and decrease in the decomposition of soil organic matter and heterotrophic respiration in the soil. However, there have been contradictory findings on the effects of nitrogen fertilisation on the decomposition of soil organic matter and soil respiration (e.g. Magill et al. 2004; Frey et al. 2014; Högberg et al. 2017). The maintenance of higher stocking in thinnings, together with longer rotations, may also increase the annual mean carbon sequestration and carbon stock in forest ecosystems over a rotation period (Liski et al. 2001; Routa et al. 2019). Overall, carbon sequestration and storage may be increased in forests in different ways by modifying current forest management practices. However, the same measures may affect forests differently, as outlined in Table 6.2. Also,

Table 6.2 Possible measures to increase carbon sequestration and storage in forests over a stand rotation

Measures at stand level	Carbon sequestration	Carbon storage
Use of improved, more productive and climate-adapted forest regeneration material	+	+
Proper region–/site-specific cultivation of different tree species	+	+
Use of mixed-species stands	+/−	−
Maintenance of higher stocking in thinning	+	+
Use of fertilisation	+	+/−
Use of longer rotation	−	+
Use of shorter rotation in storm-, drought-, fire-, insect- or fungus-prone forests	+/−	−
Decreased drainage (low-productivity peatlands)	+/−	+/−
No management	+/−	+

+ increase, − decrease, +/− direction of effect uncertain

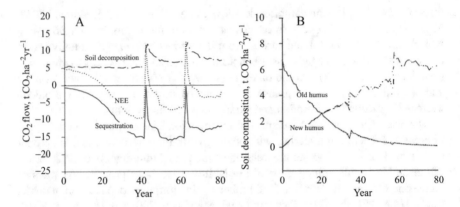

Fig. 6.1 Development of annual carbon flows of NEE (carbon sequestration + soil decomposition) (**a**) and soil decomposition of new and old humus (**b**) in a boreal Norway spruce stand after a clearcut over an 80-year rotation period with two thinnings at ages 40 and 60 years in southern Finland. Redrawn from Kilpeläinen et al. (2011). Positive values denote carbon flowing to the atmosphere, negative values denote carbon flowing to the ecosystem

management effects on the economic profitability of forest production should be considered in practical forestry.

Figure 6.1 provides an example of the development of the net ecosystem CO_2 exchange (NEE) of a boreal, even- aged Norway spruce stand on a medium-fertility upland site over an 80-year rotation period, based on gap-type forest-ecosystem model SIMA (Kellomäki et al. 2008) simulations (Kilpeläinen et al. 2011). Seedling stands (2000 seedlings ha^{-1}) act as a carbon source over the first 20 years after a clearcut because the carbon sequestration is lower in young seedling stands than the carbon emissions from decaying humus and litter in the soil. As carbon sequestration increases, a stand becomes a carbon sink. The mean annual carbon uptake over 80 years is 11.4 t CO_2 ha^{-2} year^{-1}, with the carbon emissions being 7.3 t CO_2 ha^{-2} year^{-1}. The thinnings at ages 40 and 60 years produce peaks in the carbon emissions due to harvesting and the decay of logging residuals.

In Fig. 6.2, a simulated example of the development of NEE (Fig. 6.2a) and carbon stocks (Fig. 6.2b) in a forest ecosystem is demonstrated under business-as-usual (baseline) thinning, 20% higher and lower tree stocking compared to the baseline, and an unmanaged (unthinned) boreal Norway spruce stand (Alam et al. 2017) over two rotation periods (i.e. 160 years). Over the whole 160-year period, the stands sequestered more carbon than they released (Fig. 6.2a). The NEE was the highest under higher stocking and the lowest under lower stocking. The increased carbon sequestration led to a 17% larger mean carbon stock (in the trees and soil) than in the baseline thinning, while decreased stocking led to a 21% lower carbon stock than in the baseline thinning. The mean carbon stock over the simulation period was the largest under the unmanaged regime (445 and 197 t CO_2 ha^{-1} in the trees and soil, respectively), while the mean carbon stock in the trees in the baseline thinning was 191 t CO_2 ha^{-1}, and in the soil, 111 t CO_2 ha^{-1} (Fig. 6.2b).

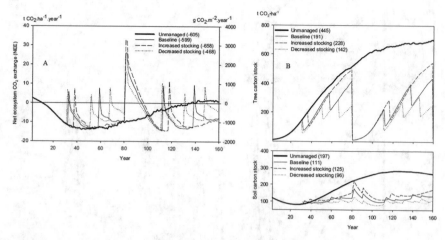

Fig. 6.2 (**a**) NEE under different management regimes in a Norway spruce stand under boreal conditions over a 160-year period under different management regimes. (**b**) Development of ecosystem carbon stocks (expressed as CO_2) in trees (top) and soil (bottom) under different management regimes. Values in parentheses in the legends indicate mean NEE (**a**) and mean carbon stock (**b**) over the simulation period. Each reduction in the tree carbon stock corresponds to the harvesting of timber from the ecosystem and its mobilisation to the technosphere as harvested wood products. After Alam et al. (2017)

Figure 6.3 shows an example of how alternative forest management regimes (use of better-growing seedlings, nitrogen fertilisation, higher stocking in thinning) might increase the simulated NEE of a forest ecosystem under even-aged management in a boreal upland Norway spruce stand with (BT, basic thinning) and without (BT-NO BIO, no bioenergy harvesting) harvesting logging residues from a clearcut. The highest increases, compared to BT-NO BIO, were observed with the use of improved seedlings in regeneration (i.e. 20% better growth than seedlings of forest-seed origin) and nitrogen fertilisation (2–4 times during a rotation period at the same time as thinning, depending on the management regime), along with the maintenance of (30%) higher stocking in thinnings over a rotation compared to the baseline management. The increases in NEE in these regimes, compared to BT-NO BIO, varied between 22 and 200%. Maintaining a higher growing stock over the rotation also increased the carbon benefits when compared to BT.

6.3   Impacts of Management and Harvesting Intensity on Carbon Storage in Forests

Forest resources comprise mosaics of single stands with varying climatic and site conditions and forest structures (age, tree species composition and stocking), which together affect the future of carbon sinks and storage in forests, and the forest harvesting potential, in different regions (Hudiburg et al. 2009; Kilpeläinen et al. 2017;

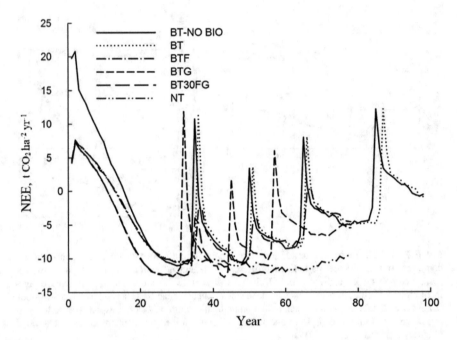

Fig. 6.3 Annual NEE (t CO_2 ha^{-1} year^{-1}) of a Norway spruce stand under alternative management regimes, with harvesting of logging residues, stumps and coarse roots (BT, BTF, BTG and BT30FG, NT) and baseline forest management (BT-NO BIO), with no harvesting of logging residues. *F* nitrogen fertilisation, *G* use of genotypes with 20% increased growth, *BT30* use of 30% higher stocking in thinnings, *NT* no thinning. (After Kilpeläinen et al. 2016)

Thom et al. 2018). At the regional level, the development of carbon sequestration and carbon storage in forests is strongly affected by the initial age structure of the forests, which also affects possible management measures over time (Baul et al. 2020). Therefore, differences in past forest management regimes in European countries will also reflect the future potential of increased carbon sequestration, wood production and carbon stocks in forests.

Heinonen et al. (2017) showed that, with around 73 million m³ of annual timber harvesting, the carbon storage of Finnish forests (forest biomass and soil), excluding forest conservation areas, may remain quite stable over the 90-year simulation period, compared to a situation with the initial growing stock (Fig. 6.4). However, with a lower even-flow timber harvest, the 40–60 million m³ levels may increase significantly. On the other hand, despite the harvesting level, the forest carbon stock starts to decrease after the first 40 years of the simulation period due to the changing forest age structure. This decline is also relatively greater at a lower harvesting intensity, which is associated with a larger share of unmanaged forests with decreasing growth and increasing mortality over time. Heinonen et al. (2017) did not consider either the effects of intensified forest management or climate change on the forest growth, or natural disturbances. By intensifying forest management, for example, by using improved regeneration materials and nitrogen fertilisation on

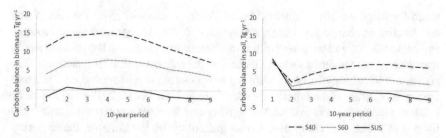

Fig. 6.4 Development of the carbon balance (i.e. the difference between sequestrated and released carbon) in the forest biomass and soil in three cutting scenarios in Finland, for nine 10-year periods under current climate. S40 and S60 denote cutting scenarios with 40 and 60 million m³ year⁻¹ cutting drains, respectively. In the SUS (sustainable) cutting scenario, the cutting drain was the highest possible (73 million m³ year⁻¹), which it was assumed would not lead to decreasing growing stock volume during the 90-year period without assuming improved forest management or climate change. Redrawn from Heinonen et al. (2017)

Table 6.3 Possible measures to increase carbon sequestration and storage in European forests and thus mitigate climate change

Measures at regional (national) level	Carbon sequestration	Carbon storage
Increase forest growth by different measures	+	+
Reduce harvesting level	+	+/−
Increase forest conservation area	+	+/−
Reduce disturbance risks in storm-, drought-, fire- or insect-prone forests by considering risk in adaptive management	+	+
Reduce deforestation and increase afforestation and reforestation	+	+

+ increase, − decrease, +/− direction of effects uncertain

upland forest sites, both the growing (carbon) stock and wood production could increase under boreal conditions with minor climate change (e.g. the RCP2.6 forcing scenario) in the coming decades (Heinonen et al. 2018a, b).

In European forests, the carbon storage (and sink) could be increased by modifying current forest management practices and harvesting intensities. However, same measures may affect the carbon sequestration and storage in different ways, especially over different time periods (Table 6.3).

When seeking to enhance the carbon storage in forests, it is important to bear in mind that forest disturbances are likely to increase in the future, with changing climate (Seidl et al. 2014; Venäläinen et al. 2020). Given this, the risk of decreasing forest carbon storage might increase, and therefore appropriate adaptation measures would be required to minimise the harmful effects (see also Chap. 5). The severity of climate change will also affect the carbon dynamics of forests through its effects on forest regeneration, growth and mortality processes, as controlled by management. These effects may also be contradictory, depending on the region.

Forests also contribute to climate through the absorption or reflection of solar radiation, cooling as a result of evapotranspiration, and the production of cloud-forming aerosols (Kalliokoski et al. 2020). These will affect the role of forests in climate change mitigation. An increase in the aboveground forest biomass and carbon stock, and the proportion of coniferous tree species in the growing stock, may decrease the planet's surface albedo (i.e. the reflection of solar radiation). This may result in enhanced climate warming in opposition to a lower carbon stock and greater proportion of broadleaf tree species (e.g. Lukeš et al. 2013). On the other hand, in managed, even- and uneven-aged boreal Norway spruce stands with relatively low average stocking over a management cycle, for example, the opposing effects on radiative forcing of changes in the albedo and carbon stocks may largely cancel each other out, providing few remaining net climate remediation benefits (Kellomäki et al. 2021). Alternatively, the maintenance of higher ecosystem carbon stocks in managed forests, or with no management, clearly implies greater net cooling benefits. This is despite the lower albedo enhancing radiative absorption, and thus enhancing warming. However, increasing the use of the no-management option may require compensation for forest owners for lost harvest income (Kellomäki et al. 2021).

Under sustainable forest management, the impacts of that management on ecosystem services other than carbon sequestration and its storage in forests, such as the production of timber and non-wood products, the maintenance of biodiversity and recreational value, should also be considered. This is important because carbon storage in forests and the amount of deadwood (an indicator of biodiversity), for example, correlate positively with each other, but negatively with harvested timber volume and the economic profitability of forestry (Diaz et al. 2021). Lower management and harvesting intensities will also lead to forest structures in which there are more older trees, a larger share of broadleaves and a greater amount of deadwood compared to forests under higher management and harvesting intensities (Heinonen et al. 2017).

6.4   Uncertainties Associated with Future Carbon Storage and Sinks

The future development of the carbon storage and sinks in European forests will be affected by the intensity of forest management and harvesting (and thus wood demand), the severity of the climate change (Kindermann et al. 2013) and the associated increase in natural forest disturbances (Seidl et al. 2014) in the different regions. In addition, the demand for multiple ecosystem services and different national and international strategies and policies (e.g. European Green Deal, climate and biodiversity policies) will affect the intensity of forest management and harvesting in those different regions. Thus, due to the complexity of the issue, there are many uncertainties in the future development of carbon sequestration and storage in European forests, and their ability to contribute to climate change mitigation.

In the EU Reference Scenario (EC 2016), forest harvests are projected to increase by 9% between 2005 (516 million m^3) and 2030 (565 million m^3) due to a growing

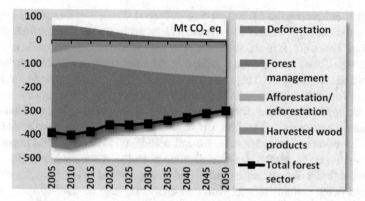

Fig. 6.5 Development of the EU-28's emissions/removals in the forest sector in Mt. CO_2eq. up to 2050 (EC 2016)

demand for energy biomass and material use (Fig. 6.5). Consequently, forest growth is projected to decrease by 3%, and the carbon sink in forests by 32%, by 2030. This may be partially compensated for by increasing the carbon sink through afforestation and decreasing emissions from deforestation. In 2050, total forest growth is, however, clearly predicted to be higher than the wood harvests in this Reference Scenario (Fig. 6.5). Assuming a constant harvest scenario (e.g. Pilli et al. 2017), the carbon sinks in the forest pools of the EU-28 are estimated to decrease by 6% in 2030 compared to the average of the historical period 2000–2029. On the other hand, based on projections for forest resources under alternative management and policy assumptions, the increased carbon storage in the EU-28 forests could provide additional sequestration benefits of approximately up to 172 Mt CO_2 year^{-1} by 2050 (Nabuurs et al. 2017).

With the right set of incentives in place at the EU and Member States levels, the EU has the potential to achieve an additional mitigation impact of 441 Mt CO_2 year^{-1} by 2050 (Nabuurs et al. 2017). The measures to achieve this would include improving forest management, expanding the forested area (afforestation), substituting for fossil- based materials and energy by wood, and setting aside forest reserves for short-term carbon sequestration. In addition to mitigating GHG emissions, the suggested measures could also adapt and build forest resilience, sustainably increase forest productivity and incomes, and tackle multiple policy goals set for the future (see also Chap. 9).

6.5   Research Implications

Changes in the intensity of forest management and harvesting will affect the carbon sequestration potential of forests and the carbon storage in forests. Using different forest management measures could help to increase these. However, it should be noted that enhancing the carbon storage in forests through management may also increase the effects of natural forest disturbances, such as wind storms, fires, drought and pests. Therefore, it is crucial to consider how to increase forest resilience in the EU through forest management. Adapting thinning regimes, shortening rotation periods and using improved regeneration materials may help to decrease the vulnerability of forests to various natural disturbances, as well as providing the means for maintaining and enhancing forest carbon sinks. It should also be considered how and under what conditions various silvicultural methods, such as stand density control, fertilisation and mixed-species forests, could help to maintain and improve the adaptation capacity, resilience and mitigation potential of forests in parallel.

Besides forest carbon sequestration and storage, wood-based products can provide significant carbon storage. Wood products may also be used to substitute for fossil-fuel-intensive materials, products and energy (Nabuurs et al. 2017; Leskinen et al. 2018). However, regional conditions vary significantly across the EU. This partly explains the difficulties involved in quantifying the mitigation impacts of the EU-level forests and the forest-based sector. Moreover, the large diversity of abiotic and biotic circumstances and management practices also makes it challenging to generalise the results of individual studies to the EU level. On the other hand, variations in the growth potential and forest utilisation rates in the various value chains create a wide range of options for adaptation to, and mitigation of, climate change in the EU, depending on regional conditions. Beyond adaptation and mitigation, the simultaneous provisioning of multiple ecosystem services for society should also be ensured, in a sustainable way, while increasing forest resilience to natural disturbances. This requires thought to be given to the uncertainties associated with climate change and the risks in forest-management decision-making, which are still understudied topics, requiring further input.

6.6 Key Messages

- European forests have acted as carbon sinks for the last few decades due to increases in the forest area, improved forest management and changing environmental conditions.
- The future development of carbon sequestration and storage in European forests will be affected both by the intensity of forest management and harvesting (associated with future wood demand) and the severity of climate change and the related increase in natural disturbances.
- The great diversity of abiotic and biotic circumstances, management practices and forest utilisation levels in the different regions of the EU creates both a wide range of options, but also challenges, for the adaptation to, and mitigation of, climate change in different regions.

References

Alam A, Strandman H, Kellomäki S, Kilpeläinen A (2017) Estimating net climate impacts of forest biomass production and utilization in fossil-intensive material and energy substitution. Can J For Res 47:1010–1020

Baul TK, Alam A, Strandman H, Seppälä J, Peltola H, Kilpeläinen A (2020) Radiative forcing of forest biomass production and use under different thinning regimes and initial age structures of a Norway spruce forest landscape. Can J For Res 50(6):523–532. https://doi.org/10.1139/cjfr-2019-0286

Bergh J, Nilsson U, Allen HL, Johansson U, Fahlvik N (2014) Long-term responses of Scots pine and Norway spruce stands in Sweden to repeated fertilization and thinning. For Ecol Manag 320:118–158

Díaz-Yáñez O, Pukkala T, Packalen P, Lexer M, Peltola H (2021) Multi-objective forestry increases the production of ecosystem services. Forestry 94(3):386–394. https://doi.org/10.1093/forestry/cpaa041

Etzold S, Ferretti M, Reinds GJ, Solberg S, Gessler A, Waldner P, Schaub M, Simpson D, Benham S, Hansen K, Ingerslev M, Jonard M, Karlsson PE, Lindroos A-J, Marchetto A, Manninger M, Meesenburg H, Merilä P, Nöjd P, Rautio P, Sanders TGM, Seidling W, Skudnik M, Thimonier A, Verstraeten A, Vesterdal L, Vejpustkova M, de Vries W (2020) Nitrogen deposition is the most important environmental driver of growth of pure, even-aged and managed European forests. For Ecol Manag 458:117762. https://doi.org/10.1016/j.foreco.2019.117762

European Commission (2016) EU reference scenario 2016. Energy, transport and GHG emissions. Trends to 2050. European Commission, Directorate-General for Energy, Directorate-General for Climate Action and Directorate-General for Mobility and Transport. https://ec.europa.eu/energy/sites/ener/files/documents/20160713%20draft_publication_REF2016_v13.pdf. Accessed 4th Jan 2021

European Environment Agency (2016). European forest ecosystems, state and trends. EEA Report 5/2016. https://doi.org/10.2800/964893

European Environment Agency (2017) Forest: growing stock, increment and fellings. https://www.eea.europa.eu/data-and-maps/indicators/forest-growing-stock-increment-and-fellings-3/assessment. Accessed 9 Jun 2020

Eurostat (2020) Agriculture, forestry and fishery statistics – 2019th edn. https://doi.org/10.2785/798761

Forest Europe (2015) State of Europe's forests 2015 report. Forest Europe, Bonn

Forest Europe (2020) State of Europe's forests 2020. Forest Europe, Bratislava

Frey SD, Ollinger S, Nadelhoffer K, Bowden R, Brzostek E, Burton A, Caldwell BA, Crow S, Goodale CL, Grandy AS, Finzi A, Kramer MG, Lajtha K, LeMoine J, Martin M, McDowell WH, Minocha R, Sadowsky JJ, Templer PH, Wickings K (2014) Chronic nitrogen additions suppress decomposition and sequester soil carbon in temperate forests. Biogeochemistry 121(2):305–316. https://doi.org/10.1007/s10533-014-0004-0

Gold S, Korotkov A, Sasse V (2006) The development of European forest resources, 1950 to 2000. For Policy Econ 8:183–192. https://doi.org/10.1016/j.forpol.2004.07.002

Gundersen P, Thybring EE, Nord-Larsen T, Vesterdal L, Nadelhoffer KJ, Johannsen VK (2021) Old-growth forest carbon sinks overestimated. Nature 591:E21–E23. https://doi.org/10.1038/s41586-021-03266-z

Haapanen M, Janson G, Nielsen UB, Steffenrem A, Stener LG (2015) The status of tree breeding and its potential for improving biomass production – a review of breeding activities and genetic gains in Scandinavia and Finland. Skogsforsk, Uppsala

Heinonen T, Pukkala T, Mehtätalo L, Asikainen A, Kangas J, Peltola H (2017) Scenario analyses for the effects of harvesting intensity on development of forest resources, timber supply, carbon balance and biodiversity of Finnish forestry. For Policy Econ 80:80–98. https://doi.org/10.1016/j.forpol.2017.03.011

Heinonen T, Pukkala T, Asikainen A, Peltola H (2018a) Scenario analyses on the effects of fertilization, improved regeneration material, and ditch network maintenance on timber production of Finnish forests. Eur J For Res 137:93–107

Heinonen T, Pukkala T, Kellomäki S, Strandman H, Asikainen A, Venäläinen A, Peltola H (2018b) Effects of forest management and harvesting intensity on the timber supply from Finnish forests in a changing climate. Can J For Res 48:1124–1134

Hengeveld GM, Nabuurs G-J, Didion M, van den Wyngaert I, Clerkx APPM, Schelhaas M-J (2012) A forest management map of European forests. Ecol Soc 17(4). https://doi.org/10.2307/26269226

Högberg P, Näsholm T, Franklin O, Högberg MN (2017) Tamm review: on the nature of the nitrogen limitation to plant growth in Fennos candian boreal forests. For Ecol Manag 403:161–185

Hudiburg T, Law B, Turner DP, Campbell J, Donato D, Duane M (2009) Carbon dynamics of Oregon and Northern California forests and potential land-based carbon storage. Ecol Appl 19:163–180. https://doi.org/10.1890/07-2006.1

Hynynen J, Salminen H, Ahtikoski A, Huuskonen S, Ojansuu R, Siipilehto J, Lehtonen M, Eerikäinen K (2015) Long-term impacts of forest management on biomass supply and forest resource development: a scenario analysis for Finland. Eur J For Res 134:415–431

Kalliokoski T, Bäck J, Boy M, Kulmala M, Kuusinen N, Mäkelä A, Minkkinen K, Minunno F, Paasonen P, Peltoniemi M, Taipale D, Valsta L, Vanhatalo A, Zhou L, Zhou P, Berninger F (2020) Mitigation impact of different harvest scenarios of Finnish forests that account for albedo, aerosols, and trade-offs of carbon sequestration and avoided emissions. Front For Glob Change 3:1–112. https://doi.org/10.3389/ffgc.2020.562044

Kellomäki S (2017) Managing boreal forests in the context of climate change. In: Impacts, adaptation and climate change mitigation. CRC Press, Boca Raton

Kellomäki S, Peltola H, Nuutinen T, Korhonen KT, Strandman H (2008) Sensitivity of managed boreal forests in Finland to climate change, with implications for adaptive management. Phil Trans R Soc B 363:2339–2349. https://doi.org/10.1098/rstb.2007.2204

Kellomäki S, Väisänen H, Kirschbaum MUF, Kirsikka-Aho S, Peltola H (2021) Effects of different management options of Norway spruce on radiative forcing through changes in carbon stocks and albedo. Forestry cpab010. https://doi.org/10.1093/forestry/cpab010

Kilpeläinen A, Strandman H, Alam A, Kellomäki S (2011) Life cycle assessment tool for estimating net CO_2 exchange of forest production. Glob Change Biol Bioenergy 3(6):461–471

Kilpeläinen A, Alam A, Torssonen P, Ruusuvuori H, Kellomäki S, Peltola H (2016) Effects of intensive forest management on net climate impact of energy biomass utilisation from final felling of Norway spruce. Biomass Bioenergy 87:1–8

Kilpeläinen A, Strandman H, Grönholm T, Ikonen V-P, Torssonen P, Kellomäki S, Peltola H (2017) Effects of initial age structure of managed Norway spruce forest area on net climate impact of using forest biomass for energy. Bioenergy Res 10:499–508. https://doi.org/10.1007/s12155-017-9821-z

Kindermann GE, Schörghuber S, Linkosalo T, Sanchez A, Rammer W, Seidl R, Lexer MJ (2013) Potential stocks and increments of woody biomass in the European Union under different management and climate scenarios. Carbon Balance Manag 8:2

Koehl M, Hildebrandt R, Olschofsky K, Köhler R, Rötzer T, Mette T, Pretzsch H, Köthke M, Dieter M, Abiy M, Makeschin F, Kenter B (2010) Combating the effects of climatic change on forests by mitigation strategies. Carbon Balance Manag 5:8

Leskinen P, Cardellini G, González-Garcia S, Hurmekoski E, Sathre R, Seppälä J, Smyth C, Stern T, Verkerk PJ (2018) Substitution effects of wood-based products in climate change mitigation. From science to policy 7. European Forest Institute, Joensuu

Lindner M, Fitzgerald JB, Zimmermann NE, Reyer C, Delzon S, van der Maaten E, Schelhaas M-J, Lasch P, Eggers J, van der Maaten-Theunissen M, Suckow F, Psomas A, Poulter B, Hanewinkel M (2014) Climate change and European forests: what do we know, what are the uncertainties, and what are the implications for forest management? J Environ Manag 146:69–83. https://doi.org/10.1016/j.jenvman.2014.07.030

Liski J, Pussinen A, Pingoud K, Mäkipää R, Karjalainen T (2001) Which rotation length is favourable to carbon sequestration? Can J For Res 31:2004–2013. https://doi.org/10.1139/x01-140

Lukeš P, Stenberg P, Rautiainen M (2013) Relationship between forest density and albedo in the boreal zone. Ecol Model 261–262:74–79

Lundmark T, Poudel BC, Stål G, Nordin A, Sonesson J (2018) Carbon balance in production forestry in relation to rotation length. Can J For Res 48(6):672–678

Luyssaert S, Schulze E-D, Börner A, Knohl A, Hessenmöller D, Law BE, Ciais P, Grace J (2008) Old-growth forests as global carbon sinks. Nature 455:213–215

Magill AH, Aber JD, Currie WS, Nadelhoffer KJ, Martin ME, McDowell WH, Melillo JM, Steudler P (2004) Ecosystem response to 15 years of chronic nitrogen additions at the Harvard Forest LTER, Massachusetts, USA. For Ecol Manag 196(1):7–28. https://doi.org/10.1016/j.foreco.2004.03.033

Moomaw WR, Law BE, Goetz SJ (2020) Focus on the role of forests and soils in meeting climate change mitigation goals: summary. Environ Res Lett 15(4):045009. https://doi.org/10.1088/1748-9326/ab6b38

Nabuurs G-J, Lindner M, Verkerk PJ, Gunia K, Deda P, Michalak R, Grassi G (2013) First signs of carbon sink saturation in European forest biomass. Nat. Clim Chang 3(9). https://doi.org/10.1038/NCLIMATE1853

Nabuurs G-J, Delacote P, Ellison D, Hanewinkel M, Lindner M, Nesbit M, Ollikainen M, Savaresi A (2015) A new role for forests and the forest sector in the EU post-2020 climate targets. From science to policy. European Forest Institute. doi:https://doi.org/10.36333/fs02

Nabuurs G-J, Delacote P, Ellison D, Hanewinkel M, Hetemäki L, Lindner M (2017) By 2050 the mitigation effects of EU forests could nearly double through climate smart forestry. Forests 8:484. https://doi.org/10.3390/f8120484

Nilsen P (2001) Fertilization experiments on forest mineral soils: a review of the Norwegian results. Scand J For Res 16:541–554

Olsson P, Linder S, Giesler R, Högberg P (2005) Fertilization of boreal forest reduces both autotrophic and heterotrophic soil respiration. Glob Change Biol 11(10):1745–1753. https://doi.org/10.1111/j.1365-2486.2005.001033.x

Pilli R, Grassi G, Kurz WA, Fiorese G, Cescatti A (2017) The European forest sector: past and future carbon budget and fluxes under different management scenarios. Biogeosciences 14:2387–2405. https://doi.org/10.5194/bg-14-2387-2017

Pretzsch H, Biber P, Schütze G, Uhl E, Rötzer T (2014) Forest stand growth dynamics in Central Europe have accelerated since 1870. Nat Commun 5:4967. https://doi.org/10.1038/ncomms5967

Routa J, Kilpeläinen A, Ikonen V-P, Asikainen A, Venäläinen A, Peltola H (2019) Effects of intensified silviculture on timber production and its economic profitability in boreal Norway spruce and Scots pine stands under changing climatic conditions. Forestry 92(3):1–11. https://doi.org/10.1093/forestry/cpz043

Rytter L, Ingerslev M, Kilpeläinen A, Torssonen P, Lazdina D, Löf M, Madsen P, Muiste P, Stener L-G (2016) Increased forest biomass production in the Nordic and Baltic countries – a review on current and future opportunities. Silva Fenn 50(5):1660

Saarsalmi A, Mälkönen E (2001) Forest fertilization research in Finland: a literature review. Scand J For Res 16:514–535

Seidl R, Schelhaas MJ, Rammer V, Verkerk PJ (2014) Increasing forest disturbances in Europe and their impact on carbon storage. Nat Clim Chang 4(9):806–810

Spiecker H (2003) Silvicultural management in maintaining biodiversity and resistance of forests in Europe—temperate zone. J Environ Manag 67(1):55–65. https://doi.org/10.1016/S0301-4797(02)00188-3

Thom D, Rammer W, Garstenauer R, Seidl R (2018) Legacies of past land use have a stronger effect on forest carbon exchange than future climate change in a temperate forest landscape. Biogeosciences 15:5699–5713

Venäläinen A, Lehtonen I, Laapas M, Ruosteenoja K, Tikkanen O-P, Viiri H, Ikonen V-P, Peltola H (2020) Climate change induces multiple risks to boreal forests and forestry in Finland: a literature review. Glob Change Biol 26:4178–4196. https://doi.org/10.1111/gcb.15183

Verkerk P-J, Fitzgerald JB, Datta P, Dees M, Hengeveld GM, Lindner M, Zudin S (2019) Spatial distribution of the potential forest biomass availability in Europe. For Ecosyst 6:5. https://doi.org/10.1186/s40663-019-0163-5

Vilén T, Cienciala E, Schelhaas MJ, Verkerk PJ, Lindner M, Peltola H (2016) Increasing carbon sinks in European forests: effects of afforestation and changes in mean growing stock volume. Forestry 89:82–90. https://doi.org/10.1093/forestry/cpv034

Chapter 7
Contribution of Wood-Based Products to Climate Change Mitigation

Elias Hurmekoski, Jyri Seppälä, Antti Kilpeläinen, and Janni Kunttu

Abstract Forest-based products—often referred to as harvested-wood products (HWPs)—can influence the climate through two separate mechanisms. Firstly, when wood is harvested from forests, the carbon contained in the wood is stored in the HWP for months to decades. If the amount of wood entering the market exceeds the amount of wood being discarded annually, this can lead to a HWP sink impact. Secondly, HWPs typically have a lower fossil carbon footprint than alternative products, so, for example, using wood in construction can lower fossil emissions by reducing the production of cement and steel, resulting in a substitution impact. The international greenhouse gas (GHG) reporting conventions and the related Intergovernmental Panel on Climate Change guidance covers the HWP sink impact, but not the substitution impacts. The HWP sink impact is restricted to tracing biogenic carbon flows, whereas the substitution impact typically covers fossil carbon flows exclusively. Importantly, the substitution and HWP sink impacts do not represent the climate- change mitigation impact of wood use, as such. Instead, they are important pieces of the broader puzzle of GHG flows related to the forest sector. This chapter presents the state-of-the-art approaches for determining the HWP sink and substitution impacts, and concludes with the policy and research implications.

Keywords Carbon storage · Climate-change mitigation · Displacement factor · Harvested wood products (HWPs) · Life-cycle assessment · Substitution impact

E. Hurmekoski (✉)
University of Helsinki, Helsinki, Finland
e-mail: elias.hurmekoski@helsinki.fi

J. Seppälä
Finnish Environment Institute, Helsinki, Finland

A. Kilpeläinen
University of Eastern Finland, Joensuu, Finland

J. Kunttu
European Forest Institute, Joensuu, Finland

© The Author(s) 2022
L. Hetemäki et al. (eds.), *Forest Bioeconomy and Climate Change*, Managing
Forest Ecosystems 42, https://doi.org/10.1007/978-3-030-99206-4_7

7.1 Contribution of Wood Products to Climate-Change Mitigation

In harvesting, woody biomass is transferred from the ecosystem to the *techno-sphere*. Wood is harvested to meet various needs, such as construction, energy, hygiene and communication.

Forests and forest-based products have a wide range of impacts across the economy and the environment, and can therefore contribute to the United Nations (UN) Sustainable Development Goals in several ways. The renewability of wood resources can aid in improving resource efficiency when substituting for plastics, for example. Favouring wood- based textiles in place of cotton-based textiles reduces the need for fresh water for irrigation and obviates the need for pesticides, while releasing land for afforestation or food production. Wood-based industrial prefabrication practices can reduce the noise and dust pollution associated with construction. Bio-based chemicals can reduce the eco-toxicity and human toxicity of commodities. Such impacts can be captured using standardised life-cycle assessment (LCA) methods. Besides a wider range of benefits, a more comprehensive analysis could also reveal possible trade-offs, such as between climate and biodiversity, or between greenhouse gas (GHG) emissions and the livelihoods of small-scale entrepreneurs, such as cotton farmers. However, to keep the scope manageable, in this chapter, we focus only on the climate-change mitigation aspect of forest-based products.

Forest-based products or energy may compete with products or energy carriers made from alternative materials. *Substitute products* can be defined as those products that provide interchangeable value or service in terms of economic utility or technical function. Printed newspaper, wood-based textiles or carton board packaging serve as examples of substitutes—they may be consumed in place of digital media, cotton or plastic, respectively. Some forest products have no apparent substitutes, such as toilet tissue, and thereby no competition, except for water.

Replacing products on the market can exert impacts on the climate due to the different emissions intensities of the substitute products. There is uncontested evidence that wood-based products are, on average, associated with lower fossil- based GHG emissions compared to non-wood products or energy carriers (Sathre and O'Connor 2010; Leskinen et al. 2018). In other words, by using wood products in place of more fossil-emission-intensive materials, greater production-.related fossil-based emissions can be avoided, thus avoiding the accumulation of additional atmospheric carbon from the use of fossil resources. Thus, in the context of forest-based climate-change mitigation, the term *substitution impact* refers to the amount of fossil emissions avoided when using wood-based products or energy in place of alternative products or energy carriers.

Besides the substitution impacts, wood products can contribute to climate-change mitigation by storing carbon in products for extended periods of time, which can lead to a sink effect, typically referred to as *harvested-wood product* (HWP) *carbon storage*. HWPs[1] act as temporary storage for the bio-based carbon seques-trated by trees from the atmosphere. The impact of this carbon flow on net emissions depends on the level of harvest and the products produced from wood. That is, the *HWP pool* acts as a carbon sink when input to the product pool exceeds outflow from the product pool (i.e. the product pool is increasing). In contrast, if the change in the overall HWP carbon stock is negative, the HWP pool acts as a source of emissions. In the GHG inventory reporting rules under the Paris Agreement, the HWP sink impact of all HWPs manufactured in the producer countries is attributed to producer countries, regardless of export destination (UN Framework Convention on Climate Change [UNFCCC] 2018).

Unlike the carbon sinks represented by forests and HWPs, the substitution impacts do not exist in the national inventory submissions for *GHG reporting* under the *UNFCCC*, and they are thereby not a part of the *Intergovernmental Panel on Climate Change* (IPCC) methodology for reporting emissions and removals. Although this makes the substitution impacts invisible (Holmgren 2019), one can argue that they are implicitly included in the form of reduced emissions in other sectors, such as construction. However, the producer countries cannot therefore directly benefit from the substitution impacts, as this would effectively lead to double counting. That is, the producer would gain substitution credits explicitly, even though these credits would already have been implicitly accounted for in the countries where the production of more emissions-intensive products were lowered as a consequence of substitution, either in the export destinations of the HWPs or in third-party countries that would have exported steel, for example. These substitution impacts can only be calculated and interpreted against a separately determined reference, and are therefore not necessarily directly comparable to the absolute reported emissions and sinks.

Substitution and HWP sinks can be a part of (national) climate policy, such as when promoting wood construction in a government programme (e.g. Finnish Government 2019), but should not be viewed in isolation from other climate-change mitigation strategies (see Chap. 8). Despite their abstract nature, the substitution and HWP sink impacts form an important part in the overall carbon flows associated with forests and wood use to and from the atmosphere.

Climate-change mitigation measures are always forward-looking. Regardless of the current situation, only additional measures compared to a baseline ought to be regarded as mitigation. Thus, it is necessary to distinguish between the substitution impacts and HWP carbon sinks originating from the current use of wood and the possible changes in the use of wood. The former gives an estimate of the amount of emissions that would occur if non-wood products were used in place of HWPs—that is, the already achieved mitigation. Only the latter (i.e. a marginal increase in the use of wood) can possibly be attributed to further efforts on climate-change mitigation. However, substitution can occur both ways. A reduction in the market share

---

[1] HWPs are synonymous with forest-based products. The term HWP has been established in technical and policy nomenclature, despite it seeming somewhat illogical.

of wood from the baseline may increase fossil emissions if, for example, wood was replaced by coal in the energy sector or concrete in the construction sector.

Despite the fairly intuitive basic principles, quantifying the substitution and carbon-storage impacts is technically complex and demanding. The results are also highly dependent on the applied system boundaries and other assumptions. In this chapter, we first introduce the basic approaches for quantifying HWP substitution and sink impacts, and then draw attention to, and discuss, their potential pitfalls.

7.1.1 Product-Level Substitution Impacts

The estimation of substitution impacts caused by an HWP replacing a specific product is the first step in estimating the total substitution impacts of wood utilisation. In practise, the substitution impacts of HWPs are calculated with the help of product-specific *displacement factors* (DFs). A DF measures how many units of fossil GHG emissions are avoided when using one unit of HWP in place of a specific alternative product. For example, if the DF for a product were 1 tCO_2eq/m^3, this would mean that using 1 m^3 of a wood-based product in a certain end use would avoid 1 t of carbon-dioxide-equivalent emissions.

The estimation of DFs is based on GHG data obtained from LCAs of HWPs and their non-wood-product counterparts. The methodology for LCAs has been standardised (Finkbeiner et al. 2006), and there are guidelines for calculating LCAs (Joint Research Centre 2010; PAS 2050:2011 2011; EN 15804:2012 2012). A necessary requirement for the comparison of items is that pairs of wood-based and non-wood-based products must have the same *functional units*, such as 1 m^2 of a building with the same functionality in terms of energy efficiency, for example. The functional unit provides a reference against which the inputs (raw materials and land use) and outputs (emissions) are calculated. The results of such calculations vary depending on the data quality, system boundaries and assumptions used in the life-cycle analyses that quantify the fossil GHG emissions of the compared products using the same functional unit. Therefore, the resulting DF estimates are often product specific.

Formally, the DF for product *i* is typically defined as:

$$DF_i = \frac{GHG_{alternative} - GHG_{wood}}{WU_{wood} - WU_{alternative}}, \qquad (7.1)$$

where $GHG_{alternative}$ and GHG_{wood} are the fossil GHG emissions resulting from the use of the non-wood and wood alternatives, expressed in mass units of carbon derived from CO_2 equivalents in a timeframe of 100 years, and WUwood. and $WU_{alternative}$ are the amounts of wood used in the wood and non-wood alternatives, expressed in mass units of carbon contained in the wood (Sathre and O'Connor 2010). Using standardised carbon units results in a unitless ratio (tonne C/tonne C), which makes the DFs comparable across widely varying cases. A positive DF value represents reduced emissions, where an alternative product is replaced by a wood-based product, while a negative value stands for the opposite.

The DF ought to be disaggregated to separately assess the impacts of different life-cycle stages—the production, use, cascading (reuse or recycle) and disposal stages (Leskinen et al. 2018). It may be possible to attribute several life-cycle stages to one DF, but doing so should be made explicit to avoid double counting.

Regarding the divisor in Eq. (7.1), the DFs can be estimated for either the amount of wood (carbon) required to produce a wood-based product or the amount of wood (carbon) contained in the final product. When upscaling the product -level substitution impacts to a region or a market, the latter approach is more straightforward to apply, as this allows allocating the correct DF to the correct feedstock flow while avoiding double counting (see Sect. 7.1.2).

Calculating DFs requires making a number of assumptions. For example, for textiles, it is more straightforward than for a construction product, as one can assume 1 t of viscose to provide roughly the same function as 1 t of cotton or polyphenols, while for construction products, one needs to consider at least the widely varying densities of the materials and the design of the functions they provide in terms of load-bearing capacity, service life, energy efficiency, fire load, etc.

Although DFs are specified for single-product pairs, their values can still range significantly from one estimate to another. Also, there is large variation in DFs between products and product categories (Table 7.1). Perhaps the most reliable average DF values are for construction and energy, due to the relatively large number of cases assessed having relatively converging estimates. For example, for chemicals, the DFs can vary significantly due to a very large number of possible combinations of feedstock, pretreatment options, sugars, conversion technologies, downstream processes, as well as end uses (Taylor et al. 2015).

Around half of all harvested wood is used for energy, including wood harvested directly from forests or as sidestreams from and byproducts of wood harvesting and industrial operations. When looking at emissions factors per unit of energy created, the emissions from biomass burning are higher than those from burning fossil fuels (e.g. Zanchi et al. 2012). As Table 7.1 suggests, bioenergy nevertheless results in positive substitution impacts due to the DF capturing only fossil emissions, while

Table 7.1 Average displacement factors for broad product categories found in the literature

Product category	Average substitution impact for wood products (tC/tC)
Structural construction products	1.3
Non-structural construction products (e.g. window panes, doors, flooring, cladding)	1.6
Textiles	2.8
Other (e.g. chemicals, furniture, packaging)	1–1.5
Average across all product categories	1.2
Energy	0.75

Soimakallio et al. (2016) for energy, Leskinen et al. (2018) for products the biogenic emissions are accounted for in the land use, land-use change and for-estry (LULUCF) sector (see Chap. 8 and Sect. 7.1.2). For the same reason, the cli-mate impact of different bioenergy fractions may differ substantially, even if not reflected in the DF. That is, while the DF will be different depending on which energy source the woody biomass is assumed to be substituting for, the release of biogenic emissions in a counterfactual situation depends on, for example, whether the biomass fraction is solid wood or byproducts and residues that would decay more quickly if left in forests and not used for energy (e.g. Repo et al. 2012).

Based on a comprehensive review of studies that have estimated DFs for wood-based products, the average DF for all included life-cycle stages has been estimated to be 1.2 tC/tC across a wide range of wood products and substitute products (Leskinen et al. 2018). This compares to an average of 2.1 tC/tC, determined previ-ously by Sathre and O'Connor (2010). Although there can indeed be systematic changes through time in the average DFs, the apparent difference between these two average estimates may be partly incidental, but may also reflect differing scopes for determining the DFs, such as in terms of separating biogenic and fossil carbon flows.

Importantly, the average substitution impact found in the literature (1.2–2.1 tC/tC) should not be used as such, as it does not have a meaningful interpretation, for at least two reasons. Firstly, the value is an arithmetic average of all those cases that have been assessed in the literature so far, but ignores those wood uses that have not been assigned a DF. Assessing an average DF for an entire market is more compli-cated than defining a DF for a single pair of products. To estimate a weighted DF for the overall wood use, it would be necessary to determine all wood flows related to forest- based products and all of the alternative products that the forest-based prod-ucts are substituting for. Taking the average DF reported in the literature and multi-plying it by the total wood use to estimate the overall substitution impacts of wood use would ignore the wood flows and those wood uses that have no substitution impacts, therefore yielding a flawed result. Secondly, these DFs capture only the difference in fossil-based emissions, while the *biogenic carbon emissions* are counted as changes in the carbon stocks in forests and HWPs, which adds an impor-tant aspect of time dynamics due to the circulation of carbon in ecosystems versus permanent fossil emissions. The static product-level DFs therefore do not capture the full impact of using wood on the climate over time, but do provide an important piece of the overall calculation framework (see Chap. 8). In the following, we exam-ine the process of upscaling substitution impacts from the product level to the mar-ket level.

7.1.2   *Market-Level Substitution Impacts*

There is no single, established way of deriving market-level substitution impact estimates. Besides assumptions, the approaches can differ in terms of the scope of the upscaling exercise. Similarly to the different approaches used for calculating the HWP sink impact (Rüter et al. 2019), the substitution impacts can be defined using production data, consumption data or value-chain data (Knauf 2015). However, the differences, in terms of calculation routines, are minor.

Deriving product-level substitution impact estimates requires having data on: (i) the volume of demand (in mass units of carbon, tC) for a product or energy carrier; (ii) the end-use distribution of the product (share of all end uses, %); (iii) the mix of substitute products in the end market (market share, %); and (iv) a DF (tC/tC) for all substitution cases (a product replacing another product in a specific end use). Taking dissolving pulp, primarily used for textile fibres, as an example, Table 7.2 shows which data are needed to derive a volume-weighted DF for an intermediate product. These assumptions are very streamlined, as the case ignores other fibres in the market, such as traditional wool and, more importantly, the emerging recycled fibres of various origins. The case also assumes that all viscose and lyocell products perform the exact same function as the fibre they are substituting for, which is unlikely to be the case in reality. The products may have different qualities, such as moisture-wicking properties, proneness to wrinkles, ability to be washed at different temperatures, etc. Moreover, the average energy mix and production structure differs between regions, which means different DFs for different regions; that is, the DFs are not only product specific, but also region specific. Lastly, defining a weighted DF for the end uses of dissolving pulp other than textiles would require an extensive survey, as there are dozens of end uses, with dissolving pulp displacing possibly more than one other material in each end use. Probably, the overall weighted DF for dissolving pulp could therefore be higher than the case suggests, as a quarter of the volume, with dozens of end uses, is ignored.

If the products were more complex than in the case of textile fibres, extensive background analysis would be required to define all possible substitution cases and their respective DFs. For example, different structural solutions in construction (light frame versus massive frame) can produce a tenfold difference in the relative wood-use intensity (cubic metres of wood used per square metre of a building) to gain the same functionality, which ought to be mirrored in the respective DF estimates to avoid large errors. There are also differences, for example, between the end-use distributions of different wood species (Poljatschenko and Valsta 2021). Moreover, there are opposite points of view—whether wood-based energy ought to

Table 7.2 Hypothetical example of the data required to derive a displacement factor (tC/tC) for dissolving pulp

End use/share		Specific product/end use		Displacing what/ share		DF (production stage)	Weighted DF
Textiles	75%	Viscose	50%	Polyolefins	75%	1.0	1.11
				Cotton	25%	−0.28	
		Lyocell	50%	Polyolefins	75%	2.59	
				Cotton	25%	1.3	
Other (explosives, detergents, sausage skins, etc.)	25%			–		–	

be assumed to substitute for an average energy mix or for fossil fuels only. Again, the assumptions in such cases ought to be made case by case, as the marginal energy sources may differ, for example, in heat production and power production, and may vary between regions.

Once the data (i–iv) have been gathered and the necessary assumptions made, the production-stage substitution impact (PSI) for a single product or energy carrier, i, can be calculated as:

$$PSI_i(t) = DF_i(t) \times S_i(t), \qquad (7.2)$$

where DFi is the volume-weighted DF for the avoided fossil-based GHG emissions (expressed as tC) per carbon contained in product i (tC), S_i is the annual volume of a wood product produced (MtC year^{-1}) and t is year. For example, taking the annual production of dissolving pulp in the world in 2018 (8.4 Mt) and multiplying it by the weighted DF from the example in Table 7.2 (1.11 tC/tC), an estimate of avoided emissions of 9.3 MtCO$_2$eq/year is derived.

A similar exercise could be performed for all relevant life-cycle stages. For example, the energy recovery of wood- based products at their end-of-life can yield substitution impacts. The end-of-life substitution impact (ESI) for product i are determined as:

$$ESI_i(t) = DF\_EoL_i(t) \times OF_i(t), \qquad (7.3)$$

where DF_EoL_i is the avoided fossil-based carbon emissions (tC) per carbon contained in product i (tC) for the end-of-life stage (incineration), and OF is the outflow from the HWP pool (MtC year^{-1}); that is, the volume of wood products accumulated over decades of historical wood harvesting exiting the HWP pool as the wood products in use are gradually discarded. For example, assuming that the annual outflow of products based on dissolving pulp from the HWP pool was the same as the annual production in 2018 (8.4 Mt) (i.e. assuming a steady state HWP pool), and multiplying this volume by a DF for energy (0.7 tC/tC), avoided emissions of 5.9 MtCO$_2$eq/ year would be estimated. Together, the avoided fossil emissions from the production and end-of-life stages would, in this hypothetical case, amount to 15.2 MtCO$_2$eq/year.

Note that, without a separately defined reference for the interpretation, this overall estimate refers to the amount of avoided fossil emissions compared to a hypothetical situation in which no wood would be used, and cannot therefore be directly compared, for example, to the absolute HWP sink impacts that portray the changes in carbon stocks in a distinct time period. To make the substitution impact comparable, it is necessary to calculate the impact of a marginal change in the system compared to a counterfactual scenario (see Chap. 8); that is, to focus on the additional substitution impacts.

To derive the substitution impacts for the total wood use in a given region and time period, data (i–iv) should be gathered for each current or emerging wood-based product and the product they are substituting for. If the different biomass streams for the production and end-of-life stage substitution impacts are disregarded, the average DF of overall wood use can be calculated by summing up the substitution impacts of all products and dividing the sum by the amount of carbon contained in the total harvested biomass delivered to the technosphere. Detailed data on wood flows is required for this task, in order to, for example, estimate the material losses in harvesting and wood processing, and the share of wood-based energy used in the production of wood-based products.

Importantly, the resulting substitution impact should not, as such, be considered to be the climate-change mitigation potential of wood-based products due to the separation of biogenic and fossil emissions. That is, the relative benefit of wood-based products compared to alternative products (a positive DF value) is mostly a consequence of tracking only fossil emissions and not biogenic emissions (Rüter et al. 2016). The energy used in the production of HWPs originates, to a great extent, from wood residues. Following the IPCC methodology, bioenergy is calculated as carbon neutral (zero emissions) in energy production because biogenic emissions are considered in full in the LULUCF sector. In effect, harvesting wood reduces forest carbon sinks for a certain amount of time, which depends on regional circumstances, such as the forest growth rate and management regime. Even though, under sustainable forest management, biogenic carbon can be assumed to circulate between the ecosystem, the technosystem and the atmosphere, the carbon payback time can be so long—up to a century—that it needs to be considered in climate policy (see Chap. 8). While there can be other analytical approaches for dealing with the separation of biogenic and fossil emissions, excluding biogenic emissions from the DFs avoids double counting when assessing the net emissions of the forest-based sector. On the other hand, this highlights the role of tracing the exact wood flows for the entire wood-use system. To avoid double counting, substitution impacts should not be allocated to the sidestream flows going into the internal bioenergy use of forest-based product mills. However, energy production that is not consumed in the production process of wood-based products causes substitution impacts similar to those of the product use. Despite the importance of understanding the implications of the assumption concerning carbon-neutral bioenergy, the positive DF values can also partly be explained by the lower embodied energy of HWPs (less energy needed in their production) and the end-of-life energy recovery of wood-based products (substituting fossil energy when a HWP is incinerated at its end-of-life).

Clearly, deriving the substitution impacts for the total wood use is a daunting task. Consequently, substitution analyses have so far typically focused only on well-known, large-volume markets and have made several simplifying assumptions to keep the analysis manageable (Holmgren 2019). Moreover, as market-level substitution analyses tend to be forward- looking, one should consider how the wood-products markets, as well as the markets of the competing products, evolve over time. As demonstrated in Chap. 4, there are several plausible pathways for the evolving uses of wood. However, in practical terms, it can be impossible to reliably determine some of the DFs for future markets, as it is not possible to trace the evolving emissions profiles of novel wood-based products, nor those of the competing products. Clearly, the accuracy of the estimates is limited by the complexity of the market and the consequent lack of data, which calls for careful documentation and sensitivity analysis.

Given the challenges in the implementation and interpretation of overall substitution impact estimates, it may be more fruitful to assess mitigation scenarios with varying wood-use structures, and compare these scenarios against a reference (see, e.g., Brunet-Navarro et al. 2021), rather than focus on perfecting a single-point estimate. While this is more relevant for decision-making, it adds another layer of complexity due to the necessity to address various dynamic, and partly indirect, market responses (e.g. Howard et al. 2021).

7.1.3   Carbon Sinks of Harvested-Wood Products

Besides substitution, HWPs can contribute to climate-change mitigation by storing carbon in biomass for a certain time period, from months to centuries. The reason why the carbon stocks of HWPs are taken into account in the GHG calculations is related to the rules of the national GHG emissions inventory, determined by the UNFCCC process. According to that process, the carbon associated with wood biomass harvested and delivered to the technosphere is considered as emissions, as it decreases the carbon stock in forests. To fine-tune this streamlined assumption, by considering the extended lifetime of biogenic carbon in the technosphere, the carbon flows in products should be monitored in the annual carbon balance calculations.

An increase in the HWP carbon storage is assigned a negative value when reporting the net emissions of the forest- based sector, which refers to a sink effect. However, this is only true in terms of notation and should not be confused with the net ecosystem production (see Chap. 6). The carbon that has been sequestered by living trees in the biomass is transferred to the HWPs and stored there for a certain time. Thus, the carbon storage itself does not imply mitigation. A mitigation impact is achieved when there is an increase in the HWP carbon stock in the technosphere (i.e. when the input to the product pool is greater than the outflow from the product pool). If there is no change in the volume of the harvest, nor in the product portfolios, there is no change in the HWP carbon stock, in which case the stock change (sink impact) remains zero. As with substitution, the HWP stock change can occur both ways—if the HWP carbon stock is reduced, the HWP pool turns into a source of carbon. Whether this results in net benefits or losses for the climate over a certain time period is determined by simultaneously assessing the entire scope of carbon pools and flows in the forest sector through time (see Chap. 8).

If the initial carbon pool of HWPs is zero, or already saturated, the mitigation impact of HWP carbon storage can be increased either by increasing the level of harvesting or by increasing the relative share of wood-based products with long lifespans, such as wood construction products, in the total harvest. The net emissions balances of these two strategies are not necessarily equal in terms of climate-change mitigation over the next few decades. That is, depending on the forest growth conditions and the product portfolio, an increase in the level of harvesting can reduce the carbon sink of forests for a longer time period than the carbon can remain stored in

HWPs, on average (Heinonen et al. 2017). However, any such conclusions ought to be assessed by comparing the impacts against a holistic counterfactual scenario (see Chap. 8) due to various indirect and cascade impacts.

Compared to substitution, the HWP sink is more straightforward to estimate. This is partly because it is included in the international GHG reporting guidance by the IPCC (Rüter et al. 2019). However, it should be noted that the four approaches detailed in the GHG reporting guidance (stock change, production, atmospheric flow, simple decay) have differences in terms of their conceptual frameworks and system boundaries (Rüter et al. 2019). The system boundary may cross national borders, such as when applying the 'production' approach, in which the producing country reports carbon stock changes from HWPs produced by that country, regardless of where the HWPs are consumed and used (Rüter et al. 2019). Thus, the policy processes have been aimed at agreeing on the use of a common method for all parties to avoid double counting across national GHG inventories in order to facilitate the global stocktake of climate-change mitigation measures.

Despite there being more peer reviewed literature on the HWP sink impact than on the substitution impact, there are still considerable uncertainties in the assumptions related to HWP sink estimates. Notably, the product half-lives are generally assumed to vary between 0 (e.g. bioenergy) and 35 (e.g. sawnwood), but they remain very difficult to assess reliably (e.g. Iordan et al. 2018). While the HWP sink impact is currently marginal compared to the substitution impact, this could change in the future due to the expected average decline in the substitution impacts and the simultaneously expected increase in the *cascade use of wood* (i.e. the reuse or recycling of wood), which would extend the lifetime of the carbon in the technosphere before it being released back to the atmosphere.

7.2 Scale and Future Outlook of Substitution Impacts and Harvested-Wood Product Sinks

7.2.1 Scale of Substitution Impacts and Sinks of Wood-Based Products

Before jumping to the estimates of overall substitution and HWP sink impacts, some caveats must be re-emphasised. Importantly, the absolute substitution impact values alone should not be interpreted as the climate benefits of wood use or as the climate-change mitigation potential. Instead, they are components of the overall carbon flow; they do not necessarily provide a meaningful interpretation in isolation from other parts of the studied system and without comparison to a common reference. To guide managerial or policy decision-making, it is necessary to calculate the overall substitution impacts of wood use and use this information in calculating the *net GHG emissions of the forest-based sector* under different forest management regimes and market structures (see Chap. 8).

There are very few systematic analyses of the overall substitution impacts at the global or European levels that depict the scale of fossil emissions avoided compared to no wood use. At the global level, Roe et al. (2019) estimated that the avoided emissions potential from increasing the demand of wood products to replace construction materials ranged from 0.25 to 1 $GtCO_2eq/year$. At the European level, Holmgren (2020) estimated that the currently avoided emissions from the industrial and energy uses of wood account for -410 $MtCO_2eq/year$ (not considering the fossil emissions of the forest-based value chains of 51 $MtCO_2eq/year$). In Finland, the current annual substitution impacts of forest-based-sector activities have been estimated as accounting for between 16.6 and 35 $MtCO_2eq$, with a domestic harvest level of 65–70 Mm^3 (Soimakallio et al. 2016; Alarotu et al. 2020; Hurmekoski et al. 2020). According to the National Inventory Report submissions under the UNFCCC, the HWP sink impact in the EU was -40.6 $MtCO_2eq/year$ in 2017, whilst in Finland, it has varied between -6.6 and 1.6 $MtCO_2eq/year$ since 1990. Future research may be able to provide more detailed estimates and a more comprehensive geographical context.

The absolute carbon pools related to forests are highly sensitive to the level of harvesting, which can mean large annual fluctuations. One option to alleviate this issue, in the context of substitution impacts, is to focus on the average values across total wood use (tC/tC). Mirroring the overall shortage of substitution estimates, there are also not many studies globally that have approximated the weighted DF for overall wood use on a national level. These estimates range between 0.3 and 1.2 tC/tC in different scenarios and with varying geographical and product scopes, with an average of around 0.5 tC/tC (e.g. Werner et al. 2010; Braun et al. 2016; Suter et al. 2017; Kayo et al. 2018; Hurmekoski et al. 2020). It can be seen that the national-level average substitution impact can be smaller than the average DF reported from meta- analyses (1.2–2.1 tC/tC) (Sathre and O'Connor 2010; Leskinen et al. 2018), which makes it all the more important to keep these values separate and to understand the reason behind the difference—the scope of the wood flows and product portfolios considered.

7.2.2   Future Trends for Substitution Impacts and Harvested-Wood Product Sinks

Future estimates of substitution impacts are uncertain, not only because of the long time frames, per se, but more because of the ongoing structural changes in the forest-based industries and their possibly evolving competitive positions. Some of the new wood-based products may have superior environmental performance compared to the current state, such as alternative solvent processes for regenerated cellulose fibres for textiles (Rüter et al. 2016). This could increase the average substitution impact of wood use if produced in place of the declining communication-paper market, to which no significant substitution impact can be attributed (Achachlouei and Moberg 2015).

At the same time, we should expect an opposite trend that will diminish the average substitution impacts of wood use, such as emissions reductions in the energy and industrial sectors to comply with the Paris Agreement target (e.g. Harmon 2019). That is, these alternative products tend to be more energy intensive than wood-based products, as indicated by positive DF values. When the average emissions from energy production are lowered, the emissions reductions of the competing products will be relatively greater than for wood-based products, thereby diminishing the relative benefit of wood use. Besides the emissions from energy production, there may be large reductions in the energy intensity of production processes, as well as process-related emissions. For example, in the construction sector, the emissions from calcination in the cement production process, which currently produces around half the total emissions in cement production, could eventually be diminished (e.g. Licht et al. 2012). In the chemicals and biofuels sector, CO_2 could be captured and used as a feedstock (e.g. Kruus and Hakala 2017), which could entirely change the logic of substitution impact estimates, if applied on a large scale. Thus, when fossil emissions are eventually phased out, there will be no fossil emissions to be avoided, leading to a zero substitution impact potential, regardless of wood use.

In addition to direct emissions, the importance of recycling has been recognised in the European Union Circular Economy action plan, which aims to improve resource efficiency by keeping the value of materials, products and resources in the technological 'closed loop' system for as long as possible by, for example, reuse, recycling and product design (European Commission 2015). Technological solutions may contribute to this target by: (i) minimising the virgin feedstock demand; and (ii) improving the reuse and recycling possibilities of materials. The expected increase in the recycling rates of non-wood products may further diminish the substitution impacts of forest-based products, such as in the case of replacing recycled plastic compared to primary plastic. The impact of wood cascading on the substitution impacts remains unclear, depending on whether, for example, the recycled wood products create additional demand, or if they substitute for existing wood products. Production technologies that improve the durability, recyclability or resource efficiency of wood-based products, such as laser scanning, improved sawing techniques, waste separation technologies and recycling technologies, would increase the HWP sink.

Thus, without investments in new wood-based products with superior environmental profiles, the average substitution impacts of wood use can be expected to diminish (e.g. Keith et al. 2015). This interplay of hypothetical developments and innovations in the wood-based products sector and in competing sectors makes the outlook very uncertain. Moreover, where the wood products and competing products are produced can make a difference, as the average energy profile and the pace of emissions reductions may be different from one region to another. However, from the environmental and societal perspectives, the competition between forest-based products and alternative products is welcome, as it strengthens the incentives for developing products and processes that cause less harm for ecosystems.

While there are still huge barriers to the large-scale uptake of novel negative emissions technologies, they will be necessary in the long-term, due to the sluggish rate of global emissions reductions (IPCC 2018). Introducing negative emissions technologies, such as bioenergy with carbon capture and storage (BECCS) or biochar, in the forest value chain, could also influence the relative benefits of wood-based products. However, this dynamic would not necessarily be captured by the DFs, if they only track fossil emissions, and should therefore be reflected elsewhere in GHG inventories. This could have implications both for research and policy. Research should consider whether the use of DFs is the most appropriate approach in such cases, and international agreements should determine who would benefit from the uptake of such technologies. These decisions could possibly influence the rate of BECCS market uptake.

7.3   Role of Harvested Wood Products in Climate-Change Mitigation

Seppälä et al. (2019) introduced the concept of a *required DF*, which depicts the required scale of substitution impacts that would exceed the temporary loss of ecosystem carbon when wood is being harvested. In other words, it depicts the minimum value for the average substitution impacts that would result in an immediate net reduction of emissions, despite an increased level of harvesting. The level of the required DF that would satisfy this condition has been estimated to be between 1.9 and 2.5 tC/tC (Seppälä et al. 2019; Köhl et al. 2020). This compares to an estimated current weighted overall substitution impact of around 0.5 tC/tC in Finland (Soimakallio et al. 2016; Hurmekoski et al. 2020), suggesting that the substitution impacts alone would not be large enough to compensate for the loss of the forest carbon sink , even in the medium- and long-term. In other words, in the Finnish context, a marginal increase in the use of wood is unlikely to reduce the net carbon emissions of wood use within the timespan of a century (see also Heinonen et al. 2017). The Finnish forest sector is characterised by long rotation periods (60–120 years), an intensive forest management regime with a young-stand-dominated age class structure, a relatively low level of natural disturbances, and a pulp- and paper-dominated industry structure, which together help to explain this difference. In other regions, the circumstances may allow opposite conclusions within a reasonable time frame.

However, it is important to acknowledge that the scope of the analysis leading to these conclusions is limited. Added to the uncertainty of the substitution and carbon sink estimates of forests and their products, such analyses tend to disregard other possibly relevant determinants of net emissions, such as the possible risks and benefits associated with the impact of climate change on forests, including increased forest growth and the different abiotic and biotic damage caused to forests by windstorms, drought, insects, pathogens and forest fires. Such elements may have a decisive impact on the conclusions, although state-of-the-art research faces difficulties in capturing them all under a single framework. Moreover, mitigation strategies are naturally influenced by a holistic assessment of efficiency and feasibility, which broadens the scope from physical carbon flows to, for example, the incentives created by harvesting income to finance further mitigation measures.

Regardless of the harvest level, the net emissions of the forest-based sector can, at least in principle, be reduced by changing the process and production structure to increase resource efficiency and the share of products with very high DFs and long life-spans, in addition to increasing the carbon sequestration with the help of forestry practices. More specifically, in terms of climate-change mitigation, improvements in the wood utilisation patterns from the perspective of the climate can be divided into two general actions: (i) reducing the share of energy involved in overall wood use, and satisfying the operational energy demand of the pulp mills and sawmills through alternative, low-emissions energy sources or by increasing the energy efficiency of such mills; or (ii) improving resource efficiency in the manufacturing of products and/or applying more cascade loops to increase the length of HWP carbon storage. In general, the use of wood for energy, in most cases, produces a lower DF compared to its material uses, while new wood-based products, such as chemicals, textiles and mixed-material composites, exhibit the highest potential (Soimakallio et al. 2016; Leskinen et al. 2018). Therefore, wood-material flows, including secondary flows, such as sidestreams and waste wood, should primarily be used for those high DF applications before being used as combustion for energy. In wood construction, the substitution potential is, on average, estimated to be slightly lower, but the HWP sink is considerably higher compared to many of the new wood-based products. Thus, the assumed carbon-storage time of up to 70 years compensates for the smaller substitution impact.

A big question mark in this context concerns the extent to which the production structure could change. The market structure is simultaneously influenced by consumer demand, the competitive advantage of a firm, industry, region or country to produce a certain product, and the strategies of the industries. For example, it is unrealistic to assume that the production of short life-span products with no substitution impacts, such as hygiene papers, would come to a halt. Even if this was the case in one region, the production would likely shift elsewhere, at least in part (see Box 7.1). Nonetheless, besides reducing the direct energy use of industrial sidestreams, one clear opportunity relates to the declining demand for communication papers, and the resulting increased availability of pulpwood for alternative uses, such as packaging (Hurmekoski et al. 2018).

It is necessary to remember that, ultimately, the primary focus of any climate-change mitigation strategy ought to be on minimising overall emissions to the atmosphere rather than, for example, maximising the substitution impacts of wood use, whether this means favouring certain products or, for example, changing forest management practices for improved ecosystem resilience. Due to established industries being required to find several complementary emissions-reduction pathways to meet the obligations of the Paris Agreement, the increased use of wood becomes all the more relevant for climate-change mitigation, especially the more pessimistic the overall climate policy outlook. However, as lignocellulosic resources will, in any case, be insufficient for replacing the entire fossil-based economy, the uses of wood will need to be prioritised in those markets where significant emissions reductions seem the most difficult to obtain, or where co-benefits on, for example, the water footprint could be gained, markets allowing.

7.4   Research Implications

Estimating substitution impacts and HWP carbon storage is a technically complex task and, in practical terms, limited by data availability. Unlike for HWP sinks, there is no IPCC guidance, or other established guidance, for deriving market- level substitution estimates, although the LCA used to derive product-specific DFs is based on international standards. The complexity of the markets and carbon flows causes variation in the scope of assumptions, whilst the abstract nature of substitution and the lack of precise data cause significant uncertainty in substitution estimates. Even though the few available estimates indicate a reasonably similar scale, this may be the result of using, at least partly, the same few data sources and similar market assumptions.

The GHG data produced by LCA is a key data source for assessing the DFs of wood-based products. However, the current practise of LCA does not include GHG emissions caused by LULUCF. In order to assess the total climate impacts of wood utilisation, including changes in the carbon stocks in forests, and their products and substitution effects, the impacts of LULUCF on GHG emissions caused by alternative non-wood products should also be taken into account. For example, the production of cotton causes land-use change, releasing CO_2 emissions from soils.

DF data on the use and end-of-life stages of products remains scarce. In particular, the impact of recycling on the overall DFs is unclear for many products. More research is needed to avoid false conclusions.

Product substitution is an abstract and essentially unobservable phenomenon. Thus, a satisfactory understanding of substitution impacts cannot be gained by looking at carbon pools and flows alone—market dynamics also require consideration. While substitution can, to some extent, be traced using market shares (e.g. Batten and Johansson 1987), our overall understanding of the occurrence and nature of substitution in the wood-based products markets remains incomplete, if not fragile. For example, consider the case of deriving an end-use-weighted DF for dissolving pulp (Sect. 7.1.2). Measuring the market share may not fully capture the substitution dynamics in the market. If the production of cotton remains stable, having reached a limit (e.g. due to no more land or water being available for its production), and at the same time man-made cellulosic fibre (MMCF) consumption increases, is it then a logical interpretation that MMCFs substitute for cotton? If the production of MMCFs were not increased to meet the increased demand for textile fibres, perhaps the remainder of the demand would have been met by synthetic fibres. Thus, it is not clear whether we should assume that MMCFs will substitute for synthetic fibres or cotton. More importantly, we cannot be sure if we can assume substitution to have occurred in the first place. That is, in the absence of perfect substitutes, the overall demand for textile fibres may have remained lower due to the markets adjusting to increased prices as a consequence of constrained supply.

Even if we accept the premise of substitution between wood-based and alternative textile fibres, can we be certain that this prevents the extra fossil feedstocks from being used? Due to *carbon leakage* (Box 7.1), an additional unit of wood products consumed does not necessarily lead to a unit reduction in other consumption, as the consumption of the alternative product may shift elsewhere in the economy (e.g. Sathre and O'Connor 2010), or the use of fossil feedstocks may be delayed for a certain period (Harmon 2019). Thus, the (non-)permanence of avoided emissions can end up being an issue equally as complex as the (non-)permanence of forest carbon sinks.

Box 7.1. Carbon Leakage
The term 'carbon leakage' refers to a shift in emissions-intensive production from one region to another, for example, if a carbon tax is operationalised in one region only. This issue is not specific to any sector, but is a generic result of supply shifting to unregulated areas to meet the global demand, when the supply is restricted either directly or through pricing. Besides shifting the production capacity, leakage can also occur in the supply side, such as in forestry, for example, when forest land is set aside (e.g. for enhancing carbon sinks through a compensation scheme for forest owners or to conserve forests through the UN's Reducing Emissions from Deforestation and forest Degradation [REDD+]) schemes. In this case, the reduced supply of wood in one region may result in an increased level of harvesting elsewhere, thus watering down the initial aim. According to Pan et al. (2020), carbon leakage is estimated to be higher in the forest sector than in energy-intensive industries.

In the forest sector, the term has mostly been used in the context of limiting the level of wood harvesting and the resulting *international carbon leakage*. Kallio and Solberg (2018) suggested that a change in harvest level in one country may lead to up to a 60–100% opposite change in the rest of the world. In addition to international leakage in a sector, there may also be *intersectoral* or *intertemporal carbon leakage*. This relates to the assumptions about substitution, in that an additional unit of wood products consumed does not necessarily lead to a unit reduction in other consumption, as the consumption of the alternative product may shift from one use to another (Sathre and O'Connor 2010). That is, if wood replaces concrete in the construction sector, the avoided use of fossil feedstocks may end up being used in another sector. Similarly, substitution may delay the use of fossil feedstocks for a certain period, but not necessarily avoid its use altogether (Harmon 2019). In other words, the substitution impact would be slowing down the fossil depletion rate, but not necessarily preventing it from happening.

Unless a policy has perfect coverage over the entire world economy, it may be impossible to avoid leakages. Although the exact rate of leakage is somewhat uncertain, it should nevertheless be considered when designing climate policies, not to encourage measures that lead to suboptimal or controversial impacts. This calls for long-term and integrated land-use planning (Pan et al. 2020).

A few practical implications arise from the identified uncertainties. There is a need to balance between expected developments in the forest-products markets and the availability of data for the determination of DFs. While there can be alternative ways for addressing the structural changes in forest-sector modelling to gain market scenarios, we simply do not have data for accurately determining the DF for novel wood-based products, or their counterparts that are not yet in production. We can, however, make assumptions about the factors that will influence the outcome, and can rely on sensitivity analysis to test their impact. Indeed, in future endeavours to estimate the scale of the overall substitution impacts, a range of estimates based on minimum and maximum assumptions should be considered, rather than a single value, together with extensive sensitivity analysis on the critical uncertainties. Focusing on marginal substitution, and using optimisation techniques to define optimal substitution cases, allows for the formulation of tangible and policy- relevant strategies (Smyth et al. 2017).

Importantly, there is a need for integrated modelling frameworks to capture the various market dynamics, such as *rebound effects*. For example, Antikainen et al. (2017) found that using textiles for longer before disposal, or substituting synthetic fibres for MMCFs, increases the overall material consumption of the economy, as this drives the consumption from textiles towards other commodities, which are often more material-intensive than textiles, and because synthetic fibres are produced from the sidestreams of the oil industry. Such impacts cannot be captured by focusing on the forest- product markets alone, but require a broader understanding of the end-use markets and value chains, as well as consumer behaviour.

A further layer of complexity is added when the fossil carbon flows are compared against the biogenic carbon flows. While GHG molecules and their impact on the climate obviously cannot be told apart based on their origin, this distinction is necessary when defining long-term mitigation strategies. In the next chapter of this book, we combine the insights of Chaps. 6 and 7, highlighting the possible trade-offs, and searching for no-regret mitigation strategies in light of these carbon flows.

7.5 Key Messages

- HWPs can influence the climate through two separate mechanisms. Firstly, when wood is harvested from a forest, the carbon contained in the wood is stored in wood-based products for months to decades. If the amount of wood products entering the market exceeds the amount of wood products being discarded annually, this can lead to a HWP *sink impact*. Secondly, as wood-based products have, on average, a lower fossil carbon footprint than alternative products, using wood in construction, for example, can avoid larger fossil emissions by reducing the production of cement and steel. This is a *substitution impact*. The international GHG reporting conventions and the related IPCC guidance only cover the HWP sink impact, but not the substitution impacts.

- Substitution impacts are measured by tracking market developments (material flows, end uses, consumer demands) and the emissions profiles of wood-based products versus products they substitute for. So far, there is no established framework for upscaling the substitution impacts to the market level. While there are rough estimates of substitution impacts, it is unclear what percentage of current or future wood-based products may ultimately substitute for fossil feedstocks. Even some of the principles remain uncertain, such as the extent to which substitution can be assumed to occur in individual cases. Despite these uncertainties, HWPs can, at least in principle, further contribute to climate-change mitigation by changing the production structure of forest industries by, for example, shifting from communication-paper manufacturing to textile manufacturing, or by shifting the use of by-products from energy to material uses.
- Importantly, the substitution and HWP sink impacts do not represent the climate-change mitigation impact of wood use, as such. Instead, they are important pieces of the broader puzzle of GHG flows related to the forest sector, and need to be considered in decision-making accordingly. This is because substitution impacts only depict changes in fossil emissions, while changes in biogenic emissions are accounted for in the LULUCF sector, indicating a possible short-term trade-off between substitution impacts and HWP sink impacts on one hand, and forest carbon sinks on the other. These fossil and biogenic carbon flows need to be tracked across time and across markets in alternative scenarios compared to a common reference in order to yield relevant policy implications.

Acknowledgements Elias Hurmekoski wishes to acknowledge financial support from the SubWood Project (No. 321627), funded by the Academy of Finland.

References

Achachlouei MA, Moberg Å (2015) Life cycle assessment of a magazine, part II: a comparison of print and tablet editions. J Ind Ecol 19:590–606

Alarotu M, Pajula T, Hakala J, Harlin A (2020) Metsäteollisuuden tuotteiden ilmastovaikutukset (Climate impacts of forest industry products). VTT-CR-00682-20

Batten DF, Johansson B (1987) Substitution and technological change. In: The global forest sector: an analytical perspective. Wiley, New York, pp 278–305

Braun M, Fritz D, Weiss P, Braschel N, Büchsenmeister R, Freudenschuß A, Gschwantner T, Jandl R, Ledermann T, Neumann M (2016) A holistic assessment of greenhouse gas dynamics from forests to the effects of wood products use in Austria. Carbon Manag 7:271–283

Brunet-Navarro P, Jochheim H, Cardellini G, Richter K, Muys B (2021) Climate mitigation by energy and material substitution of wood products has an expiry date. J Clean Prod 127026

EN: 15804:2012 (2012) Sustainability of construction works. Environmental product declarations. Core rules for the product category of construction products, British Standards Institute, London

European Commission (2015) Closing the loop – an EU action plan for the circular economy. European Commission, Communication COM (2015) 614/2

Finkbeiner M, Inaba A, Tan R (2006) The new international standards for life cycle assessment: ISO 14040 and ISO 14044. Int J Life Cycle Assess 11:80–85. https://doi.org/10.1065/lca2006.02.002

Finnish Government (2019) Programme of Prime Minister Sanna Marin's Government 10 December 2019. Inclusive and competent Finland—a socially, economically and ecologically sustainable society. Publications of the Finnish Government 2019:33, Helsinki

Harmon ME (2019) Have product substitution carbon benefits been overestimated? A sensitivity analysis of key assumptions. Environ Res Lett 14:65008

Heinonen T, Pukkala T, Mehtätalo L, Asikainen A, Kangas J, Peltola H (2017) Scenario analyses for the effects of harvesting intensity on development of forest resources, timber supply, carbon balance and biodiversity of Finnish forestry. For Policy Econ 80:80–98

Holmgren P (2019) Contribution of the Swedish forestry sector to global climate efforts. Swedish Forest Industries Federation

Holmgren P (2020) Climate effects of the forest based sector in the European Union. Confederation of European Paper Industry

Howard C, Dymond CC, Griess VC, Tolkien-Spurr D, van Kooten GC (2021) Wood product carbon substitution benefits: a critical review of assumptions. Carbon Balance Manag 16:1–11

Hurmekoski E, Jonsson R, Korhonen J, Jänis J, Mäkinen M, Leskinen P, Hetemäki L (2018) Diversification of the forest-based sector: role of new products. Can J For Res 48:1417–1432. https://doi.org/10.1139/cjfr-2018-0116

Hurmekoski E, Myllyviita T, Seppälä J, Heinonen T, Kilpeläinen A, Pukkala T, Mattila T, Hetemäki L, Asikainen A, Peltola H (2020) Impact of structural changes in wood-using industries on net carbon emissions in Finland. J Ind Ecol 24:899–912

Iordan CM, Hu X, Arvesen A, Kauppi P, Cherubini F (2018) Contribution of forest wood products to negative emissions: historical comparative analysis from 1960 to 2015 in Norway, Sweden and Finland. Carbon Balance Manag 13:12

IPCC (2018). Global warming of 1.5°C. An IPCC special report on the impacts of global warming of 1.5°C above pre—industrial levels and related global greenhouse gas emission pathways, in the context of strengthening the global response to the threat of climate change. Intergovernmental Panel on Climate Change

JRC (2010) ILCD handbook: general guide for life cycle assessment—detailed guidance. European Union, Luxembourg

Kallio AMI, Solberg B (2018) Leakage of forest harvest changes in a small open economy: case Norway. Scand J For Res 33:502–510

Kayo C, Dente SMR, Aoki-Suzuki C, Tanaka D, Murakami S, Hashimoto S (2018) Environmental impact assessment of wood use in Japan through 2050 using material flow analysis and life cycle assessment. J Ind Ecol 23:635–648

Keith H, Lindenmayer D, Macintosh A, Mackey B (2015) Under what circumstances do wood products from native forests benefit climate change mitigation? PLoS One 10:e0139640

Knauf M (2015) A multi-tiered approach for assessing the forestry and wood products industries' impact on the carbon balance. Carbon Balance Manag 10:4

Köhl M, Ehrhart H-P, Knauf M, Neupane PR (2020) A viable indicator approach for assessing sustainable forest management in terms of carbon emissions and removals. Ecol Indic 111:106057

Kruus K, Hakala T (2017) The making of bioeconomy transformation. VTT Technical Research Centre of Finland

Leskinen P, Cardellini G, Gonzalez-Garcia S, Gustavsson L, Hurmekoski E, Sathre R, Seppälä J, Smyth C, Stern T, Verkerk H (2018) Substitution effects of wood-based products in climate change mitigation. From science to policy, no. 7. European Forest Institute. https://doi.org/10.36333/fs07

Licht S, Wu H, Hettige C, Wang B, Asercion J, Lau J, Stuart J (2012) STEP cement: solar thermal electrochemical production of CaO without CO2 emission. Chem Commun 48:6019–6021

Pan W, Kim M-K, Ning Z, Yang H (2020) Carbon leakage in energy/forest sectors and climate policy implications using meta-analysis. For Policy Econ 115:102161

PAS 2050:2011 (2011) Specification for the assessment of the life cycle greenhouse gas emissions of goods and services. British Standards Institute, London

Poljatschenko VAM, Valsta LT (2021) Carbon emissions displacement effect of Finnish mechanical wood products by dominant tree species in a set of wood use scenarios. Silva Fenn 55. https://doi.org/10.14214/sf.10391

Repo A, Känkänen R, Tuovinen J, Antikainen R, Tuomi M, Vanhala P, Liski J (2012) Forest bioenergy climate impact can be improved by allocating forest residue removal. GCB Bioenergy 4:202–212

Roe S, Streck C, Obersteiner M, Frank S, Griscom B, Drouet L, Fricko O, Gusti M, Harris N, Hasegawa (2019) Contribution of the land sector to a 1.5° C world. Nat Clim Chang:1–12

Rüter S, Werner F, Forsell N, Prins C, Vial E, Levet A-L (2016) ClimWood2030-Climate benefits of material substitution by forest biomass and harvested wood products: perspective 2030. Final report. Thünen report

Rüter S, Matthews RW, Lundblad M, Sato A, Hassan RA (2019) Chapter 12: harvested wood products. In: 2019 Refinement to the 2006 IPCC guidelines for National Greenhouse Gas Inventories. Intergovernmental Panel on Climate Change, pp 12.1–12.49

Sathre R, O'Connor J (2010) Meta-analysis of greenhouse gas displacement factors of wood product substitution. Environ Sci Pol 13:104–114

Seppälä J, Heinonen T, Pukkala T, Kilpeläinen A, Mattila T, Myllyviita T, Asikainen A, Peltola H (2019) Effect of increased wood harvesting and utilization on required greenhouse gas displacement factors of wood-based products and fuels. J Environ Manag 247:580–587

Smyth C, Rampley G, Lemprière TC, Schwab O, Kurz WA (2017) Estimating product and energy substitution benefits in national-scale mitigation analyses for Canada. GCB Bioenergy 9:1071–1084

Soimakallio S, Saikku L, Valsta L, Pingoud K (2016) Climate change mitigation challenge for wood utilization the case of Finland. Environ Sci Technol 50:5127–5134

Suter F, Steubing B, Hellweg S (2017) Life cycle impacts and benefits of wood along the value ch Ain: the case of Switzerland. J Ind Ecol 21:874–886

Taylor R, Nattrass L, Alberts G, Robson P, Chudziak C, Bauen A, Libelli IM, Lotti G, Prussi M, Nistri R. (2015) From the sugar platform to biofuels and biochemicals. Final report for the European Commission Directorate-General Energy N (ENER/C2/423–2012/SI2.673791)

UNFCCC (2018) Report of the conference of the parties serving as the meeting of the parties to the Paris agreement on the third part of its first session, held in Katowice from 2 to 15 December 2018. United Nations Framework Convention on Climate Change

Werner F, Taverna R, Hofer P, Thürig E, Kaufmann E (2010) National and global greenhouse gas dynamics of different forest management and wood use scenarios: a model-based assessment. Environ Sci Pol 13:72–85

Zanchi G, Pena N, Bird N (2012) Is woody bioenergy carbon neutral? A comparative assessment of emissions from consumption of woody bioenergy and fossil fuel. GCB Bioenergy 4:761–772

Chapter 8
Climate-Change Mitigation in the Forest-Based Sector: A Holistic View

Elias Hurmekoski, Antti Kilpeläinen, and Jyri Seppälä

Abstract Forests and wood use can contribute to climate-change mitigation by enhancing carbon sinks through afforestation, reforestation and improved forest management, by maintaining carbon stocks through natural or anthropogenic disturbance prevention, by increasing offsite carbon stocks, and through material and energy substitution by changing the industry production structure and enhancing resource efficiency. As forests grow fairly slowly in Europe, increasing the wood harvesting intensity decreases the carbon stocks in aboveground biomass, at least in the short to medium term (0–50 years) compared to a baseline harvest regime. The key issue is the time frame in which the decreased carbon stock in forests can be compensated for by improved forest growth resulting from improved forest management and the benefits related to wood utilisation. Thus, there is a need to address potential trade-offs between the short- to medium-term and the long-term (50+ years) net emissions. An optimal strategy needs to be tailored based also on regional specificities related to, for example, local climatic and site conditions, the state of the forests, the institutional setting and the industry structures. This chapter presents a way to assess the effectiveness of forest-sector climate-change mitigation strategies across different contexts and time horizons, combining the climate impacts of forests and the wood utilisation of the technosphere. We identify potential 'no-regret' mitigation pathways with minimum trade-offs, and conclude with the research and policy implications.

Keywords Carbon sink · Climate-change mitigation · Counterfactual sceanrio · Forest sector · Net greenhouse gas emissions · Substitution impacts · Time dynamics

E. Hurmekoski (✉)
University of Helsinki, Helsinki, Finland
e-mail: elias.hurmekoski@helsinki.fi

A. Kilpeläinen
University of Eastern Finland, Joensuu, Finland

J. Seppälä
Finnish Environment Institute, Helsinki, Finland

L. Hetemäki et al. (eds.), *Forest Bioeconomy and Climate Change*, Managing Forest Ecosystems 42, https://doi.org/10.1007/978-3-030-99206-4_8

8.1 Introduction

This chapter contains a synthesis of the insights in Chap. 6, dealing with forests, and Chap. 7, focusing on the technosphere. In this chapter, we adhere to the principle of *'what the atmosphere sees'* regarding climate change. What we mean by this refers to two aspects. Firstly, it is necessary to pay equal attention to all factors affecting the climate impacts of the forest sector; that is, to simultaneously analyse a biological ecosystem (forests), a technological system (industries) and a socioeconomic system (markets). This is imperative for the designing and monitoring of climate-change mitigation measures that ensure a *net* reduction in the atmospheric greenhouse gas (GHG) concentrations in a desired time period.

Secondly, the principle of 'what the atmosphere sees' can also refer to the *absolute GHG emissions and sinks*, in contrast to GHG emissions and sinks based on an *accounting framework* used for monitoring and policy purposes. The accounting of GHG emissions and sinks reported under national GHG inventories facilitates tracking of the impacts of mitigation measures, for example, by comparing annually reported values against a baseline. In the EU climate policy framework, forests and forest bioenergy are regulated under the *land use, land-use change and forestry (LULUCF) sector*, for example. Changes in carbon stocks in existing forests are compared with *forest reference levels*—that is, the level of carbon sink tied to the forest management regime of a historical reference period. Although the current LULUCF regulation (EU 2018) contains several flexibilities, the principle is that, if the sinks in managed forests decline below the reference level, in the accounting framework, these emissions need to be reduced elsewhere in the LULUCF sector, or in other sectors outside the EU emissions trading scheme. Thus, the LULUCF regulation aims to make the forest and land- use sector comparable to all other economic sectors in the EU climate policy, thereby emphasising the importance of short-term mitigation outcomes over the possible long-term benefits of wood use. Such accounting principles are a result of international policy processes that emphasise the short and medium term in climate-change mitigation. In this chapter, we refer to comparisons against a reference scenario (synonymous with a counterfactual scenario) to facilitate the drawing of policy implications based on the effectiveness of selected mitigation measures, but this should not be confused with internationally negotiated GHG accounting principles.

The mitigation potential of the forest-based sector can be realised through several alternative measures (e.g. Nabuurs et al. 2017a, b; St-Laurent et al. 2018; Intergovernmental Panel on Climate Change [IPCC] 2019), as summarised in Table 8.1. There is, however, an important caveat—some of the forest-based climate-change mitigation strategies are more effective on short-term climate impacts, whereas others are better for long-term impacts, and also some of the measures may be better suited for one particular regional context than another. Thus, there can be trade-offs between the measures. Moreover, in real life, forests are used for multiple purposes simultaneously, leading to mixed climate-change mitigation strategies that consider the balancing of different societal objectives and needs for forests.

Table 8.1 Selected climate-change mitigation measures related to forests and wood utilisation

Category	Type and timing of impact	Example and description
1A Increase forest area	Enhance sink: delayed impact	Afforestation and reforestation enhance forest carbon sinks
1B Maintain forest area	Reduce source: immediate impact	Avoid land-use change: reducing deforestation prevents biogenic emissions from occurring
2A Increase site-level carbon density	Enhance sink: delayed impact	Improve forest management: increasing the growth rate of forests and forest carbon sinks by, for example, using improved regeneration materials (seeds and seedlings) or forest fertilisation
2B Maintain site-level carbon density	Reduce source: immediate impact	Avoid forest degradation: for example, protect old-growth forests to maintain forest carbon stocks and promote forest conservation (and biodiversity)
3A Increase landscape-level carbon stocks	Enhance sink: delayed impact	Apply principles of sustainable forest management: enhancing forest carbon sequestration (growth) and maintaining higher stocking in thinning (possibly also longer rotations), while provisioning other ecosystem services
3B Maintain landscape-level carbon stocks	Reduce source: immediate impact	Increase forest resilience to natural disturbances: adaptation of forests and forest management to climate change, for example, by increasing the species diversity in forest stands, and forest resilience to different abiotic and biotic damage by various means (see Chap. 3)
4A Increase offsite carbon in wood products	Enhance sink: immediate impact (if meeting also 1B, 2B and 3B)	Increase the share of long-lived wood products: increasing the share of, for example, construction products in the overall wood-industry product portfolio to increase carbon storage outside the atmosphere, irrespective of the amount of wood harvested
4B Increase material and energy substitution	Reduce source: immediate impact (if meeting also 1B, 2B and 3B)	Increase the share of low-emission wood products: increasing the share of, for example, textiles in the overall wood-industry product portfolio to avoid fossil emissions, irrespective of the amount of wood harvested; increasing material efficiency and clean, non-burning energy in wood-based product chains to avoid fossil emissions through the reallocation of sidestreams

Modified after Nabuurs et al. (2007, Fig. 9.4). Note that the impacts of any strategy need to be assessed on a case-by-case basis, and in a comprehensive framework, in order to avoid oversimplified conclusions

As a rule, in terms of climate-change mitigation, increasing wood harvesting intensity decreases carbon stocks in forests compared to the baseline harvest, at least in the short to medium term (see Sect. 8.3). Thus, the effectiveness of a mitigation strategy depends on the *net emissions* (expressed as CO_2 equivalents) over time—the reduction in the carbon sink caused by harvesting, and the time by which the reduction is compensated for by the recovered forest carbon stock, the avoided fossil emissions and the carbon stored in products. How do we analyse the effectiveness of these strategies across different contexts?

8.2   Estimating the Impacts of Mitigation Strategies

Creating an understanding of the overall climate impact of the forest-based sector requires simultaneous consideration of carbon stock changes in standing trees, soil and harvested-wood products (HWPs), as well as the avoided fossil emissions from the substitution impacts of wood use. The forest carbon sink equalled −373.5 $MtCO_2eq$/year and the HWP sink equalled −40.6 $MtCO_2eq$/year in 2017 (European Economic Area [EEA] 2019). There are very few systematic estimates for substitution impacts, but according to Holmgren (2020), the material and energy substitution impact of wood use in Europe in 2018 accounted for −410 $MtCO_2eq$/year. For comparison, the total European GHG emissions (without the LULUCF sector) were 4333 $MtCO_2eq$/year in 2017 (EEA 2019). Note that in GHG inventories, negative values stand for removals from the atmosphere and positive values stand for emissions to the atmosphere. Note also that the estimate of overall substitution impacts refers to the amount of avoided fossil emissions compared to a hypothetical situation in which no wood would be used, and cannot therefore be directly compared to the absolute forest and HWP sink impacts that portray the changes in carbon stocks from one year to the next. Thus, adding the above individual impacts together (forest carbon sink, HWP sink, substitution) would provide no direct or necessarily meaningful interpretation without a comparison to a common reference.

In the context of climate-change mitigation, it is therefore essential to differentiate between the *current emissions balance* and the *changes in the emissions balance* as a result of mitigation strategies. This requires quantifying at least two scenarios, one with the current portfolio of mitigation actions and one with the new portfolio of mitigation actions. The difference between these two scenarios reveals the climate impacts of a new mitigation strategy relative to the current one. For this reason, the most important step in analysing the climate impacts of wood use is to compare the mitigation outcomes against a *counterfactual scenario* through time. Essentially, the counterfactual scenario determines how GHG emissions caused by wood utilisation would have developed over time, if the forest management and wood-use regime had not been subject to the selected set of climate-change mitigation strategies. For example, one could examine the difference in net GHG emissions over a 50-year period, if wood harvesting in the EU was increased by 15% compared to maintaining the current harvest level. Varying approaches have been used for this type of analysis. A useful starting point can be to compare alternative scenarios to a counterfactual scenario, determined as a reference or business-as-usual scenario, in which the sector would develop according to past trends or according to the most recent forecasts.

The difference in GHG emissions between baseline Scenario *b* and alternative Scenario *a* in time interval $[t0, T]$ can be calculated according to the following equation (Seppälä et al. 2019):

$$\Delta NGHGE_{b-a} = \int_{t_0}^{T} \frac{(TC_b(t) - TC_a(t) + SC_b(t) - SC_a(t) + PC_b(t) - PC_a(t)}{-(SI_b(t) - SI_a(t))} dt,$$

$$(8.1)$$

where TC is the tree carbon stock change, SC is the soil carbon stock change, PC is the product carbon stock change, SI represents the substitution impacts, and t is the year. If the result of Eq. (8.1) is negative, the mitigation potential of a strategy adopted in Scenario a is better than the mitigation potential of Scenario b in time interval $[t_0, T]$ (e.g. in the next 30 years). Thus, Eq. (8.1) allows us to compare the different outcomes of selected strategies on the cumulative GHG emissions over a certain time span. However, assessing the most appropriate time interval for interpreting the climate benefits of different wood-utilisation strategies is not straightforward (see Sect. 8.3). In practice, it is useful to assess the climate impacts of strategies both over the short and medium term (0–50 years) and in the long term (50+ years).

Peer-reviewed landscape-level studies that have determined the net climate impacts of mitigation scenarios against a counterfactual scenario for different harvesting intensities indicate a clear trade-off between short-term and long-term mitigation outcomes (Werner et al. 2010; Lundmark et al. 2014; Smyth et al. 2014; Matsumoto et al. 2016; Soimakallio et al. 2016; Gustavsson et al. 2017; Heinonen et al. 2017; Chen et al. 2018; Pingoud et al. 2018; Valade et al. 2018; Seppälä et al. 2019; Kalliokoski et al. 2020; Jonsson et al. 2021). The climate impacts are affected by the initial age structures of the studied landscapes in interaction with plausible management of the stands over time. For example, the positive effects of increased forest carbon sequestration through higher stocking of growing stock has been found to be greater for the initially young and middle-aged forest landscape, while the total climate impacts remain more sensitive to the substitution impact or timber-use efficiency than to the initial stocking (Baul et al. 2020).

Applying Eq. (8.1) may be difficult in practice, when considering the uncertainties relating to the long-term projections for carbon sinks and substitution impacts, such as the risk of sink reversals due to forest disturbances or changing product portfolios. Besides questions on the accuracy, the utility of the equation in terms of the managerial and policy implications depends on the scope of the factors considered when calculating the outcomes of the scenarios. For example, in the case where there is an anticipated increase in natural disturbances, one could recommend premature final felling to avoid even higher net emissions, whereas a more holistic strategy would additionally consider adaptation measures, such as increasing the tree species diversity of forest stands. Importantly, it is likely that there are at least some indirect, and not easily quantifiable, impacts missing from the calculation, such as carbon leakage, forest management incentives created by forest-owner revenues, and other socioeconomic cascade impacts, which calls for broader assessment and interpretation (e.g. Favero et al. 2020). Nonetheless, without systematic modelling tools and explicit comparisons between scenarios (such as in Eq. (8.1)), the results are not necessarily going to be transparent, and the meaning of the time span may be left without interpretation, or the interpretation may be overly simplistic.

8.3 Time Dynamics of Fossil and Biogenic Emissions

Regarding the comparisons between mitigation and counterfactual scenarios, it is important to understand that the impacts of mitigation strategies in given circumstances will change according to the selected time interval, among other scope considerations (see, e.g., Pingoud et al. 2012). In many studies, however, an interpretation of the results with regard to different time intervals is largely missing.

There are fundamental differences between biogenic and fossil carbon flows, even though the GHG compounds and their impact on the climate are identical. This is because the biogenic carbon in forests can be considered to be in balance between the biosphere and the atmosphere, if the original growth circumstances of the forests continue and the harvesting areas remain as forests. By contrast, fossil emissions disturb the carbon balance by adding carbon from geological stores to the atmosphere. Both carbons are removed from the atmosphere through photosynthesis and emitted to the atmosphere through respiration, decay and fires, but are also stored in plants, in the organic matter in soils and in HWPs.

According to the concept of *carbon neutrality*, the carbon emissions and sinks from a (managed) forest ecosystem are in balance over the long term (e.g. Nabuurs et al. 2017a, b). Therefore, in the long term, the use of biomass feedstock does not result in permanent increases in atmospheric CO_2 concentrations, when sustainably sourced. However, this definition of carbon neutrality should not be confused with what is agreed in the international GHG inventory reporting conventions. Despite the actual unit emissions from biomass burning exceeding those of fossil fuels (Zanchi et al. 2012), biomass burning is reported as zero emissions in the energy sector in order to avoid double counting between the energy sector and the LULUCF sector. This is because the carbon impact is already fully counted in the LULUCF sector as increased net emissions due to a reduction in carbon stocks in forest ecosystems as a result of harvesting wood. Thus, the actual impact of wood use on the net emissions of the economy needs to be assessed case by case, by tracking both the ecosystem and technosystem GHG flows through time. For example, if the average substitution impacts were increased to the extent that they almost offset a temporary decline in the carbon sink (compared to baseline), an increase in harvesting level could be interpreted as resulting in net neutral impacts in the short run, but in net mitigation benefits in the long run because permanent fossil emissions and sink saturation would have been avoided.

Forest biomass harvesting leads to a temporary decline in the forest carbon stock. The time lag for achieving net mitigation benefits through biomass utilisation can be described using two concepts—*carbon debt* and *carbon parity* (Mitchell et al. 2012). The carbon debt repayment period refers to the period between biomass harvesting and the point at which the overall GHG emissions balance of the harvest scenario (including potential avoided fossil emissions through wood utilisation and carbon stock in wood products) offsets the loss of carbon stored in the biomass at the time of harvesting. The concept of carbon parity also takes into account the accumulated ecosystem carbon that could have occurred had the harvest not taken place. This leads to the comparison of a scenario with the defined activities against a scenario without those activities—the counterfactual scenario. The repayment period depends on, for example, the latitude (boreal, temperate or tropical), biomass feedstock source (stemwood or residue), spatial scale (forest stand or landscape), type of fossil fuel replaced (coal, oil or gas) and energy usage (heating or power generation) (Geng et al. 2017), as well as the initial state of the forest, the forest growth rate and the management practices (Valade et al. 2018).

Reviews focused on wood-based bioenergy have determined that the range of parity times proposed in the literature exceeds two centuries (Lamers and Junginger 2013; Bentsen 2017). Bentsen (2017) found that the carbon debt and parity times vary mostly due to the assumptions used, and that methodological rather than eco-system- and management-related assumptions determine the findings. According to Lamers and Junginger (2013), parity times are primarily influenced by the choice and formulation of the reference scenario and the assumptions relating to fossil-fuel-displacement efficiency. Generally, in the EU forest context, harvesting trees for bioenergy has been estimated to have a parity time exceeding a century for final fellings, less than a century for thinnings, and from a few years to a few decades for forest residues (Nabuurs et al. 2017a, b; Pingoud et al. 2018). In some cases, such as when using forest residues, dead or damaged wood from natural disturbance sites, or new plantations on highly productive or marginal land, the net carbon ben-efits can be almost immediate (Lamers and Junginger 2013). The parity times have apparently been studied mostly in relation to bioenergy exclusively, so that evidence on the range of parity times that consider all major GHG flows (i.e. including mate-rial substitution impacts and HWP carbon sinks) remains limited. In one such assessment for Canada, the parity time ranged from 43 years to more than a century (Chen et al. 2018), depending on counterfactual assumptions. However, as noted by Bentsen (2017), the lack of consensus on carbon debt and parity times among researchers implies that the concept remains inadequate in itself for informing and guiding concrete policy development, with too many of the outcomes and conclu-sions relying on methodology and assumptions. Nonetheless, in the absence of bet-ter metrics, these concepts are helpful in understanding—at least conceptually—the temporal delay in climate benefits relating to an expanding bioeconomy.

Besides the temporal dynamics, it is necessary to note that the spatial scope of the analysis can also influence the conclusions. The broader the spatial context, the more policy-relevant the conclusions become. That is, compared to an analysis at the single forest stand level, an analysis at the landscape level ought to consider a more holistic range of contributing factors and interdependencies, even if this means some detail is lost. Importantly, at the forested landscape level, there is no carbon debt associated with a baseline harvest due to the mixture of stands in different developmental stages that average this out. The landscape-level analysis is also more relevant to analyses at the regional or national levels than the stand-level anal-ysis. Still, it is clear that more carbon could have accumulated in the ecosystem in the short to medium term with a lower harvest level, in the absence of natural disturbances, which is why the carbon parity period needs to be considered at all levels of analysis (Nabuurs et al. 2017a, b).

It has been estimated that global net emissions ought to be reduced at an annual rate of around 7% between 2020 and 2030 to be able to limit global warming to 1.5° (Olhoff and Christensen 2019). This roughly equals the annual net emissions reduction produced in 2020 by the global lockdown measures, which were on an unprecedented scale, resulting from the COVID-19 pandemic (Olhoff and Christensen 2020). This urgency may be in conflict with the carbon parity times of several decades associated with increased wood harvesting, although it has to be recognised that this depends on the counterfactuals that should also account for various market and ecosystem responses, for example, that current models typically ignore (see, e.g., Favero et al. 2020). Nonetheless, due to the potentially existing carbon parity period, it is necessary to track both the biogenic carbon dynamics and the fossil-based production systems over time in order to enable the designation of realistic and sustainable mitigation strategies that will not increase atmospheric carbon within a given time period, and at the same time will allow a rapid run-down of the fossil-based economy.

Importantly, although science can facilitate an understanding of the implications of different time scales, this is not sufficient for judging how the short- and long-term benefits should be appraised against one another, as this requires a value judgement. Such judgements may also get confused in climate policy with the motives of different stakeholder groups, such as the definition of sustainability (i.e. the level of human interference with nature) (Camia et al. 2021). The appraisal of short- and long-term climate-change mitigation measures also depends on the overall mix of mitigation policies and strategies that exist outside the forest sector through time. Thus, there is no conclusive view on what scale and in which time frame a temporary increase in atmospheric carbon can be tolerated in order to yield long-term benefits. For example, the precautionary principle would suggest that a temperature overshoot should be avoided, which might lead to the idea that the level of harvesting should be immediately reduced to promote higher forest carbon sink for the coming decades. However, biological sinks eventually become saturated, and may be prone to natural disturbances, unless managed and continually harvested to meet various human demands, so reducing the harvest level would ultimately cause higher permanent fossil-based emissions (IPCC 2019).

8.4   Viable Strategies for Climate-Change Mitigation in the Forest Sector

It is widely recognised that the forest-based sector can play an important role in climate-change mitigation. However, optimising between the short- and long-term benefits can be tricky (IPCC 2019). An optimal harvesting intensity, from the viewpoint of carbon sinks and the amount of wood utilised, will vary. Due to the complexity of the system, it is not possible to draw a clear line at a level of harvesting that could be characterised as (un)sustainable. Thus, there is clear motivation for seeking ways to reduce the net GHG emissions of the forest-based sector that would not lead to adverse consequences either in the short or the long term. In the following, we explore some examples of how this could be achieved.

Increasing the net carbon-sink capacity of forests can be achieved by simultaneously improving their carbon sequestration while reducing their GHG emissions, for example, in drained peatland forests. Forest fertilisation is the most effective measure for increasing the carbon sequestration of forests in boreal locations in the short term, whereas the use of improved forest regeneration material is an even more effective measure in the long term, but their combined use is the most effective (Heinonen et al. 2018). Also, on organic peatland forest soils, avoiding the unnecessary maintenance of ditches can result in lower decomposition rates in the peat layer and its attendant GHG (especially CO_2) emissions as a result of raising the water table.

According to FAOSTAT data, the EU27's share of world forest area was 3.9% in 2020. At the same time, the EU27's share of world forest industry exports was 40.8% in 2019 (worth US$100 billion). With such an intensive focus on providing forest-based products for global markets, the EU has a major opportunity to steer sustainable production and consumption. Indeed, the substitution impacts and HWP sinks of wood use could be increased without affecting the forest carbon sink via at least three channels. Firstly, by increasing the resource efficiency and reducing the carbon footprint of the current forest products in the entire value-chain relative to the current situation. Secondly, by changing the portfolio of current products. The byproducts of wood-using industries could be increasingly used to produce bio-chemicals, for example, and to satisfy the operational energy demands of pulp mills and sawmills using alternative (renewable) energy sources or by increasing the energy efficiency of such mills. Thirdly, by innovating new forest-based products with higher substitution impacts than the current forest products, and replacing the latter. Increasing the relative use of wood in the construction, textiles, packaging and chemicals markets in place of, for example, graphic papers would reduce the demand for concrete, steel, cotton, plastic and oil derivatives, and would plausibly result in reduced net emissions, *ceteris paribus*. However, even if the product portfolio could be influenced by strategies or policies, the demand for forest-based products will largely be shaped by consumer preferences, industry competitiveness and the availability of alternative products to satisfy the same needs. Moreover, the impacts of changes in the product portfolio ought to be assessed case by case, and considering the possible indirect impacts. Targeting an increase in the share of long-lived wood products does not guarantee climate benefits in itself, due to the markets adjusting to the changing supply and demand, which may lead to unwanted spill-over impacts. However, it may be possible to use industrial byproducts for construction, for example, in the form of concrete additives or walls made of nanocellulose, which might increase both the HWP sink and the substitution impacts compared to the baseline. Finally, markets will also always demand short-lived products, such as packaging, hygiene papers and textiles, and it makes sense to produce these with as low a carbon footprint as possible, which might also mean using wood-based products.

A key aspect for sustainability lies in addressing the overconsumption of natural resources, meaning that the demand for virgin raw materials—in particular, single-use, non-renewable materials—needs to be reduced. Apart from reducing consumption through carbon pricing, for example, this could be achieved by increasing recycling and reuse (circular economy, cascade use), and by increasing the resource efficiency of production (e.g. Böttcher et al. 2012). Increasing circularity (i.e. the cascading use of wood biomass) leads to a longer delay in the release to the atmosphere of the biogenic carbon that is stored in wood-based products, while also reducing the need to harvest virgin biomass. However, an increase in cascading use requires the avoidance of harmful substances in wood-based products, as these could hinder the effective recycling and reuse of these wood materials (European Commission 2018). Thus, eco-design is a key measure for improving the circularity and substitution effects of wood products for the future.

Besides mitigation strategies, it is necessary to simultaneously build forest resilience against the changing climate and increased forest disturbances, notably by moving from monoculture forests to mixed forests (see Chap. 4). This will also require the adaptation of industry production structures to accommodate the changing wood supply. According to Dugan et al. (2018), the most effective forest-sector mitigation measures are likely to be those that retain or enhance the co-benefits and ecosystem services of forests, such as biodiversity, water quality and the economy, in addition to achieving climate-change mitigation benefits. Moreover, the mitigation portfolios need to be regionally differentiated in order to be effective (e.g. Smyth et al. 2020).

8.5   Key Messages

- The climate impact of the forest-based sector value chain, from forestry to the disposal of forest-based products, should be analysed from the point of view of 'what the atmosphere sees'—that is, what is the net GHG impact on the atmosphere of changes in all product stages. The net climate impact of wood use is the sum of complex interactions between net carbon sinks in forests (tree and soil carbon sinks: see Chap. 6) and changes in the GHG emissions of the technosphere (HWP carbon sinks, substitution impacts: see Chap. 7), as well as the biophysical impacts related to forests (albedo, aerosols, black carbon: see Chaps. 3 and 6). The net impacts are influenced by the selected time frame, as well as future assumptions about markets (market structure, leakage effects), forest management regimes, the risks of carbon sink reversals (natural disturbances), etc. All these determinants ought to be assessed against a counterfactual scenario—what would the carbon balance have been if the selected mitigation strategies were not followed?

- It is difficult to simultaneously perceive the impacts of all these factors, not to mention capture their influence in quantitative modelling in a single peer-reviewed article, or even as part of a multidisciplinary research consortium. There is already significant uncertainty around the major components of the net GHG balance, primarily in the outcomes of models predicting the future forest carbon sink and the substitution impacts. Together with alternative system boundaries and widely varying assumptions, this may help us to understand why opinions based on science can differ. We simply do not know for certain what the optimal forest rotation or optimal production structure should be, considering all of the above factors. Because the scope of even state-of-the-art studies is limited, therefore not allowing the direct policy implications to be understood, attention is required when interpreting the results of such studies.
- Depending on the counterfactual, there can be a short-term trade-off between increasing the level of harvesting to increase the substitution impacts and reducing the level of harvesting to increase the net carbon sink. At the same time, all GHG emissions to the atmosphere need to be rapidly reduced, regardless of their origin. Thus, it becomes necessary to explore 'no-regret' strategies for boosting the forest-based bioeconomy. This includes developing new low-carbon innovations in the forest-based bioeconomy, improving the resource efficiency and circularity of the current bioproducts, and ensuring the vitality and resilience of forests against natural disturbances. The effectiveness of management measures also needs to be assessed in their socioeconomic context, paying particular attention to a rapid and just transition away from fossil-based industries. Thus, it becomes necessary to simultaneously consider mitigation and adaptation strategies, along with other societal goals.

Acknowledgements Elias Hurmekoksi wishes to acknowledge financial support from the SubWood Project (No. 321627), funded by the Academy of Finland.

References

Baul TK, Alam A, Strandman H, Seppälä J, Peltola H, Kilpeläinen A (2020) Radiative forcing of forest biomass production and use under different thinning regimes and initial age structures of a Norway spruce forest landscape. Can J For Res 50:523–532

Bentsen NS (2017) Carbon debt and payback time–lost in the forest? Renew Sustain Energy Rev 73:1211–1217

Böttcher H, Freibauer A, Scholz Y, Gitz V, Ciais P, Mund M, Wutzler T, Schulze E-D (2012) Setting priorities for land management to mitigate climate change. Carbon Balance Manag 7:5

Camia A, Giuntoli J, Jonsson R, Robert N, Cazzaniga NE, Jasinevičius G, Avitabile V, Grassi G, Barredo JI, Mubareka S (2021) The use of woody biomass for energy purposes in the EU. EUR 30548 EN, Publications Office of the European Union, Luxembourg. ISBN 978-92-76-27867-2. https://doi.org/10.2760/831621, JRC122719

Chen J, Ter-Mikaelian MT, Yang H, Colombo SJ (2018) Assessing the greenhouse gas effects of harvested wood products manufactured from managed forests in Canada. For Int J For Res 91:193–205

Dugan AJ, Birdsey R, Mascorro VS, Magnan M, Smyth CE, Olguin M, Kurz WA (2018) A systems approach to assess climate change mitigation options in landscapes of the United States forest sector. Carbon Balance Manag 13:13

EEA (2019) Annual European Union greenhouse gas inventory 1990–2017 and inventory report 2019. Submission under the United Nations Framework Convention on Climate Change and the Kyoto Protocol. European Environmental Agency, EEA/PUBL/2019/051

EU (2018) Regulation 2018/841 of the European Parliament and of the Council of 30 May 2018 on the inclusion of greenhouse gas emissions and removals from and use, and use change and forestry in the 2030 climate and energy framework, and amending Regulation (EU) No 5

European Commission (2018) A sustainable bioeconomy for Europe: strengthening the connection between economy, society and the environment. COM(2018) 673 final

Favero A, Daigneault A, Sohngen B (2020) Forests: carbon sequestration, biomass energy, or both? Sci Adv 6:eaay6792

Geng A, Yang H, Chen J, Hong Y (2017) Review of carbon storage function of harvested wood products and the potential of wood substitution in greenhouse gas mitigation. For Policy Econ 85:192–200

Gustavsson L, Haus S, Lundblad M, Lundström A, Ortiz CA, Sathre R, Le Truong N, Wikberg P-E (2017) Climate change effects of forestry and substitution of carbon-intensive materials and fossil fuels. Renew Sust Energ Rev 67:612–624

Heinonen T, Pukkala T, Mehtätalo L, Asikainen A, Kangas J, Peltola H (2017) Scenario analyses for the effects of harvesting intensity on development of forest resources, timber supply, carbon balance and biodiversity of Finnish forestry. For Policy Econ 80:80–98

Heinonen T, Pukkala T, Kellomäki S, Strandman H, Asikainen A, Venäläinen A, Peltola H (2018) Effects of forest management and harvesting intensity on the timber supply from Finnish forests in a changing climate. Can J For Res 48:1–11

Holmgren P (2020) Climate effects of the forest based sector in the European Union. Confederation of European Paper Industry

IPCC (2019) Climate change and land: an IPCC special report on climate change, desertification, land degradation, sustainable land management, food security, and greenhouse gas fluxes in terrestrial ecosystems

Jonsson R, Rinaldi F, Pilli R, Fiorese G, Hurmekoski E, Cazzaniga N, Robert N, Camia A (2021) Boosting the EU forest-based bioeconomy: market, climate, and employment impacts. Technol Forecast Soc Change 163:120478. https://doi.org/10.1016/j.techfore.2020.120478

Kalliokoski T, Bäck J, Boy M, Kulmala M, Kuusinen N, Mäkelä A, Minkkinen K, Minunno F, Paasonen P, Peltoniemi M (2020) Mitigation impact of different harvest scenarios of Finnish forests that account for albedo, aerosols, and trade-offs of carbon sequestration and avoided emissions. Front For Glob Chang

Lamers P, Junginger M (2013) The 'debt' is in the detail: a synthesis of recent temporal forest carbon analyses on woody biomass for energy. Biofuels Bioprod Biorefining 7:373–385

Lundmark T, Bergh J, Hofer P, Lundström A, Nordin A, Poudel BC, Sathre R, Taverna R, Werner F (2014) Potential roles of Swedish forestry in the context of climate change mitigation. Forests 5:557–578

Matsumoto M, Oka H, Mitsuda Y, Hashimoto S, Kayo C, Tsunetsugu Y, Tonosaki M (2016) Potential contributions of forestry and wood use to climate change mitigation in Japan. J For Res 21:211–222

Mitchell SR, Harmon ME, O'Connell KEB (2012) Carbon debt and carbon sequestration parity in forest bioenergy production. GCB Bioenergy 4:818–827

Nabuurs GJ, Masera O, Andrasko K, Benitez-Ponce P, Boer R, Dutschke M, Elsiddig E, Ford-Robertson J, Frumhoff P, Karjalainen T (2007) Forestry. Climate change 2007: mitigation. In: Metz B et al. (eds) Contribution of working group III to the fourth assessment report of the intergovernmental panel on climate change. Cambridge University Press, Cambridge/New York

Nabuurs G-J, Arets EJMM, Schelhaas M-J (2017a) European forests show no carbon debt, only a long parity effect. For Policy Econ 75:120–125

Nabuurs G-J, Delacote P, Ellison D, Hanewinkel M, Hetemäki L, Lindner M, Ollikainen M (2017b) By 2050 the mitigation effects of EU forests could nearly double through climate smart forestry. Forests 8:484

Olhoff A, Christensen JM (2019) Emissions gap report 2019

Olhoff A, Christensen JM (2020) Emissions gap report 2020

Pingoud K, Ekholm T, Savolainen I (2012) Global warming potential factors and warming payback time as climate indicators of forest biomass use. Mitig Adapt Strateg Glob Chang 17:369–386

Pingoud K, Ekholm T, Sievänen R, Huuskonen S, Hynynen J (2018) Trade-offs between forest carbon stocks and harvests in a steady state–a multi-criteria analysis. J Environ Manag 210:96–103

Seppälä J, Heinonen T, Pukkala T, Kilpeläinen A, Mattila T, Myllyviita T, Asikainen A, Peltola H (2019) Effect of increased wood harvesting and utilization on required greenhouse gas displacement factors of wood-based products and fuels. J Environ Manag 247:580–587

Smyth CE, Stinson G, Neilson E, Lemprière TC, Hafer M, Rampley GJ, Kurz WA (2014) Quantifying the biophysical climate change mitigation potential of Canada's forest sector. Biogeosciences 11:3515

Smyth CE, Xu Z, Lemprière TC, Kurz WA (2020) Climate change mitigation in British Columbia's forest sector: GHG reductions, costs, and environmental impacts. Carbon Balance Manag 15:1–22

Soimakallio S, Saikku L, Valsta L, Pingoud K (2016) Climate change mitigation challenge for wood utilization the case of Finland. Environ Sci Technol 50:5127–5134

St-Laurent GP, Hagerman S, Kozak R, Hoberg G (2018) Public perceptions about climate change mitigation in British Columbia's forest sector. PLoS One 13

Valade A, Luyssaert S, Vallet P, Djomo SN, Van Der Kellen IJ, Bellassen V (2018) Carbon costs and benefits of France's biomass energy production targets. Carbon Balance Manag 13:26

Werner F, Taverna R, Hofer P, Thürig E, Kaufmann E (2010) National and global greenhouse gas dynamics of different forest management and wood use scenarios: a model-based assessment. Environ Sci Pol 13:72–85

Zanchi G, Pena N, Bird N (2012) Is woody bioenergy carbon neutral? A comparative assessment of emissions from consumption of woody bioenergy and fossil fuel. GCB Bioenergy 4:761–772

Chapter 9
Climate-Smart Forestry Approach

Lauri Hetemäki and Hans Verkerk

Abstract The climate-smart forestry approach was pioneered in 2015 and has been generating increasing interest since then. It was developed as a response to the often very narrow and partial perspective on how forests and the forest-based sector can contribute to climate-change mitigation. Moreover, its basis is the understanding that, in order to effectively enhance climate mitigation, efforts should be made to find synergies and minimise trade-offs with the other ecosystem services forests provide, such as biodiversity, wood production and recreation. By doing this, greater support can be generated for climate mitigation measures. The approach acknowledges that there is no one-size-fits-all toolkit to cover all circumstances, but rather measures have to be tailored according to regional characteristics and institutions. In summary, climate-smart forestry is a holistic approach to how forests and the forest-based sector can contribute to climate-change mitigation that considers the need to adapt to climate change, while taking into account specific regional settings.

Keywords Climate smart forestry · Climate mitigation · Adaptation to climate change · Forest sinks · Substitution · Carbon storage

9.1 Background

The climate-smart forestry (CSF) approach was originated by Nabuurs et al. (2015, 2017), with further elaborations and arguments in Nabuurs et al. (2017), Kauppi et al. (2018), Jandl et al. (2018), Yousefpour et al. (2018), Bowditch et al. (2020) and Verkerk et al. (2020). The first CSF pilots were introduced in the Netherlands in 2019 (Dutch Climate Accord 2018). The CSF concept, as such, was introduced

L. Hetemäki (✉)
European Forest Institute, Joensuu, Finland

Faculty of Agriculture and Forestry, University of Helsinki, Helsinki, Finland
e-mail: lauri.a.hetemaki@helsinki.fi

H. Verkerk
European Forest Institute, Joensuu, Finland

© The Author(s) 2022
L. Hetemäki et al. (eds.), *Forest Bioeconomy and Climate Change*, Managing Forest Ecosystems 42, https://doi.org/10.1007/978-3-030-99206-4_9

earlier by the Food and Agriculture Organization (FAO) under the concept of climate-smart agriculture at the Hague Conference on Agriculture, Food Security and Climate Change in 2010 (FAO 2013). The FAO used CSF in a very broad sense, and primarily in addressing the developing countries.[1] Nabuurs et al. (2015, 2017) introduced CSF specifically in the context of the Paris Climate Agreement and the EU's land use, land-use change and forestry (LULUCF) policy, and since then, it has been further elaborated to highlight the linkages with climate-change adaptation (Verkerk et al. 2020).

The main idea of the CSF approach is expressed in the following statement: *climate-smart forestry is a holistic approach to how forests and the forest-based sector can contribute to climate-change mitigation that considers the need to adapt to climate change, while taking into account specific regional settings.* Stated like this, it may seem overly generalised and not necessarily providing any significant new insight. However, the discussions, scientific literature, policies (e.g. greenhouse gas [GHG] reporting to the United Nations Framework Convention on Climate Change [UNFCCC] and EU LULUCF) and interests around forests and the forest-based sector in the last few decades have illustrated how narrow, partial and incomplete they often are. There is a common tendency to stress only some specific aspect(s), such as forest sinks, forest product substitution and storage, mitigation or adaptation, but rarely are all these viewed simultaneously, in a holistic approach. Moreover, the discussions of, for example, LULUCF have often been technically quite demanding, while quantifying carbon sinks for LULUCF is complicated and involves many uncertainties. Against this backdrop, a holistic CSF approach, tailored to individual regional settings, is more novel and significant than it sounds.

Before going into detail on the CSF approach, it is useful first to outline the background and motivation behind how the approach came to be, and what it can offer in the future. In doing this, we base the discussion on Nabuurs et al. (2015, 2017) and Verkerk et al. (2020), in particular.

9.2 The Climate-Smart Forestry Approach: Origin and Objectives

In 2015, intensive preparations for the UNFCCC COP21 Paris meeting were being made. During this process, the European Forest Institute carried out a study to understand how European forests and the forest sector could best contribute to climate mitigation targets (Nabuurs et al. 2015). This was a pioneering study that put forward the CSF approach. The approach then went on to be further developed in Nabuurs et al. (2017), where it was used specifically to address the situation

---

[1] It may be noted that the state forests of Finland (Metsähallitus) also introduced climate-smart forestry into their operations (Vaara et al. 2018). See, also, the European Forest Institute video on CSF: https://www.youtube.com/watch?v=2wGjBKhw6U4

regarding the Paris Climate Agreement and the European Commission's (2016b) legislative proposal to incorporate GHG emissions and removals associated with LULUCF into its 2030 Climate and Energy Framework. The Climate and Energy Framework was aimed at a total emissions reduction of 40% by 2030 for all sectors combined, as part of the Paris Agreement (UN 2015; European Commission 2016a, b).

Even during the negotiations leading up to the Kyoto Protocol in 1997, the forest sector's role in climate mitigation was being discussed. However, concerns about the consequences of incorporating the existing forest sink into the climate targets resulted in the policy of imposing significant limits on the role of forests in climate-change mitigation (Ellison et al. 2014). In the EU policies, particular requirements relating to "caps", and "forest (management) reference levels" (now called forest reference levels) were introduced. This set of rules evolved into the EU LULUCF proposal (European Council 2017), which was later adapted as a regulation (European Council 2018). Nabuurs et al. (2017) raised concerns about the LULUCF proposal due to it limiting the role of forests and the forest sector in climate policy, expressing that this role could be much greater than what had been assessed in the initial impact assessment report (European Council 2016a).

Against this backdrop, Nabuurs et al. (2017) argued that the EU forest-based sector could contribute much more to climate mitigation than was the current state, and what had been conventionally understood. They also interpreted CSF as a more specific climate-focused approach under the more general Sustainable Forest Management concept (Forest Europe 1993). The key idea behind CSF is that it considers the whole value chain—from forest to wood products and energy—in climate mitigation and adaptation, with a focus on what the atmosphere 'sees', and giving less consideration to GHG reporting and accounting conventions. It contains a wide range of measures that can be applied to provide positive incentives for more firmly integrating climate objectives into the forest-based-sector framework. Consequently, Nabuurs et al. (2017) argued that CSF is more than just storing carbon in forest ecosystems—it rather builds upon three main objectives: (1) reducing and/or removing GHG emissions; (2) adapting and building forest resilience to climate change; and (3) sustainably increasing forest productivity and income.

These CSF objectives can be achieved by tailoring policy measures and actions to the regional circumstances of forest-based sectors in the EU Member States. Nabuurs et al. (2017) quantified an indicative potential mitigation impact of the EU forest-based sector by 2050 (Table 9.1), and suggested policy measures to incentivise action according to the three main CSF objectives.

The core of CSF is that it not only aims to realise climate-change mitigation, but also tries to achieve synergies and minimise trade-offs with other forest functions, such as adaptation to climate change, biodiversity conservation, ecosystem services and the bioeconomy. By reducing and/or removing GHG emissions, adapting and building forest resilience, and sustainably increasing forest productivity and income, it tackles multiple policy goals, such as many of those stressed in the UN Sustainable Development Goals. Nabuurs et al. (2017) argued that the greater the *synergies* and the fewer the *trade-offs* between climate policy and other societal and forest-related

Table 9.1 The climate-smart forestry approach potential mitigation effect in the EU *(all numbers are approximations and contain large uncertainties)*

Main category of Forest management measure	Sub-measure	Mitigation effect *(Mt CO_2/year)*
1. Improve forest management		172
	1a. Full-grown coppice	56
	1b. Enhanced productivity and improved management	38
	1c. Reduced disturbances, deforestation, drainage	35
	1d. Material substitution using wood products	43
2. Expand forest area		64
3. Substitute for energy		141
4. Establish forest reserves		64
Total		**441**

Source: Nabuurs et al. (2017)

goals, the more likely the climate objectives would be effectively implemented in practice.

9.3   Climate-Smart Forestry Measures Toolkit

To look in more detail at what types of measures CSF could include, Nabuurs et al. (2017) provided a summary. Here, we also summarise the potential measures and approximate their impacts on climate mitigation at the EU level. However, the estimates should be regarded as rough estimates indicating the potential relative scales rather than absolute and precise figures.

Although EU forests cover 40% of the land area, the scientific literature has occasionally pointed to a limited, but additional, mitigation role for EU forests on the order of 90–180 Mt. CO_2/year by 2040 (*Intergovernmental Panel on Climate Change* 2007). Nabuurs et al. (2017), however, found that, with the implementation of CSF, EU forests and the forest sector could play a much larger role. They indicated that the current annual mitigation effect of the EU forest-based sector, via contributions to the forest sink, and material and energy substitution, is on the scale of 13% of the current total EU emissions. With the right set of incentives, and through the implementation of CSF goals in the EU and Member States, Nabuurs et al. (2017) approximated (with high uncertainties) that the additional potential climate mitigation could be around 440 Mt. CO_2/year by 2050. Table 9.1 illustrates the different CSF measures and approximate magnitudes that could be implemented in the EU to increase the forest-based sector's climate-mitigation impact (Nabuurs et al. 2017). As stated above, the estimation is only indicative and not precise, and it does not, for example, estimate the mitigation potential against a baseline counterfactual scenario, the importance of which is highlighted in Chap. 8 of this book.

Projections of forest resources under alternative management and policy assumptions—derived from a number of different studies—indicate that carbon storage in existing EU forests could continue to increase, providing additional sequestration benefits of approximately up to 172 Mt. CO_2/year by 2050 (Nabuurs et al. 2017). Measures to achieve this could include the enhanced thinning of stands, leading to additional growth and higher-quality raw materials, regrowth of new species or provenances, the planting of more site-adapted species and provenances, and regeneration using faster-growing species and provenances. For example, large areas of low-productivity hardwoods, previously only used for firewood production (some 350,000 km^2 of old coppice forests), could be regenerated and replaced by more-productive mixed deciduous and coniferous forests, generating an additional sink of ~56 Mt. CO_2/year. This could be done by using new provenances better adapted to future climates, without the need for exotic species.

An increase in the productivity of forests through the above could potentially yield an addition to the forest sink of ~38 Mt. CO_2/year in the long term. Moreover, productivity growth would add ~35 million m^3 of future harvest potential to the EU's fellings of 522 million m^3, although possible trade-offs with other services would need to be considered. The long-term use of harvested-wood products (HWP) can also contribute to mitigation by substituting for the use of fossil fuels and energy-intensive materials, such as steel and concrete in the construction sector. According to Nabuurs et al. (2017), favouring wood-use in the construction sector (when carried out in synergy with the above-mentioned production increase) could potentially help avoid future emissions on the order of ~43 Mt. CO_2/year.

Emissions occur in European forests as well. Annual deforestation, as exemplified by land-use conversions to infrastructure of close to 1000 km^2/year, causes emissions of ~15 Mt. CO_2/year. Further, natural disturbances, such as bark-beetle outbreaks, windstorms and forest fires, on average, cause emissions of ~18 Mt. CO_2/year. The draining of peat soils under forests emits ~20 Mt. CO_2/year. Forest management and the improved protection of forest areas in the EU can reduce all of these emissions. In Spanish forests, for example, a more active management regime that also aims to introduce better-adapted species could significantly reduce fire risk and thus land-use change. Nabuurs et al. (2017) conservatively estimated that, if two-thirds of the above emissions could be avoided, this would reduce emissions by a further ~35 Mt. CO_2/year.

However, more importantly than focusing on the approximate and uncertain quantitative estimates of the mitigation impact (Table 9.1) is considering the types of forest management measures that could be implemented to enhance the EU's forest-based-mitigation potential. Below, we summarise a possible forest-management toolkit for enhancing climate-change mitigation (modified and extended from Nabuurs et al. 2013, p. 4). When applying it, it is essential to bear in mind that there is no one-size-fits-all toolkit that accommodates all circumstances, but rather it has to be tailored according to regional characteristics and institutions.

- Conserve high-carbon-stock densities in old forests that are not in high-disturbance-risk areas. Older forests tend to contain more deadwood and habitat niches than intensively managed forests, and this would also help benefit biodiversity, while constraining the average increment rates.
- Harvest mature forests that are at high risk of disturbance and already have low productivity. This would intensify the carbon sink only in the longer term. However, society would have to accept that forests may temporarily need to go through a net emissions phase—which they could also do without harvesting, if effected by disturbances—in order to safeguard long-term forest sinks.
- Conserve high-carbon-stock forests on sensitive sites, high-soil-carbon sites and steep slopes.
- Improve the management and protection of fire-prone forests to safeguard their carbon stocks. Also, reduce disturbance risks by moving increasingly away from monoculture forests to mixed forests. This would also tend to enhance biodiversity.
- Switch to continuous-cover forest management, if economic and forest management conditions allow. This favourably adjusts the ratio of productive to unproductive time spans in the management cycle.
- In forests primarily managed for wood production, optimise the silvicultural techniques (such as planting, tending and harvesting) to arrive at a carbon-efficient management scheme, and stimulate the recycling of forest raw materials and wood products.
- When using forest biomass for bioenergy, use forest residues, biomass from thinnings, coppice forests, sidestreams of the forest industry (sawchips, bark, black-liquor, etc.) and post-consumer wood.
- Continue afforestation and restoration schemes in Europe, particularly in less-forested parts. In addition, reduce deforestation, which would deliver immediate gains by avoiding emissions.

Reflecting on the developments since the Paris Climate Agreement, Verkerk et al. (2020) argued that CSF is a necessary, but still missing, component in strategies to decarbonise global society. The authors refined the CSF approach and focused on three mutually reinforcing components: (1) increasing carbon storage in forests and wood products, in conjunction with the provisioning of other ecosystem services; (2) enhancing forest health and resilience through adaptive forest management; and (3) using wood resources sustainably to substitute for non-renewable, carbon-intensive materials. Successful implementation of CSF would require policies that help to find the right balance between short- and long-term goals, as well as between the need for wood production, biodiversity protection and other important ecosystem services.

Verkerk et al. (2020) stressed the need to enhance global afforestation, and avoid deforestation and degradation, combine mitigation and adaptation measures in forest management, and use wood sustainably as a substitute for non-renewable carbon-intensive materials. The successful development of CSF calls for policymakers to create incentives for the investment needed to activate forest-management and finance-mitigation and -adaption measures, including protecting biodiversity and other ecosystem services. Such a development requires holistic policy frameworks and action plans that incorporate the requisite innovations, institutions, infrastructures and investments (i.e. the four 'I's in Rockström et al. 2017). According to Verkerk et al. (2020), in order to implement these, it is important to develop economic instruments, such as taxes, subsidies and public procurement, as well as introducing extended producer responsibilities, incentives for retaining value in the circular economy processes, and supporting all the initiatives in the context of greening the finances.

In order to illustrate what role CSF could play in different regions of the EU countries in further detail, and how local circumstances may impact its measures, we turn to look at four case studies—the Czech Republic, Finland, Germany and Spain.

References

Bowditch E, Santopuoli G, Binder F, del Río M, La Porta N, Kluvankova T, Lesinski J, Motta R, Pach M, Panzacchi P, Pretzsch H, Temperli C, Tonon G, Smith M, Velikova V, Weatherall A, Tognetti R (2020) What is climate-smart forestry? A definition from a multinational collaborative process focused on mountain regions of Europe. Ecosyst Serv 43:101113. https://doi.org/10.1016/j.ecoser.2020.101113

Ellison D, Lundblad M, Petersson H (2014) Reforming the EU approach to LULUCF and the climate policy framework. Environ Sci Pol 40:1–15

European Commission (2016a) Commission staff working document impact assessment accompanying the document proposal for a regulation of the European Parliament and of the Council. European Commission, Brussels

European Commission (2016b) Proposal for a regulation of the European Parliament and of the Council on the inclusion of greenhouse gas emissions. European Commission, Brussels

European Council (2017) Proposal for a regulation of the European Parliament and of the Council on the inclusion of greenhouse gas emissions and removals from land use, land use change and forestry into the 2030 climate and energy framework and amending regulation no 525/2013 of the European Parliament and the Council on a mechanism for monitoring and reporting greenhouse gas emissions. European Council, Brussels

Food and Agriculture Organization (FAO) (2013) Climate-smart agriculture: sourcebook. FAO of the United Nations, Rome. ISBN 978-92-5-107720-7. http://www.fao.org/3/a-i3325e.pdf. 10 Dec Jan 2020

Forest Europe (1993) RESOLUTION H1: General Guidelines for the Sustainable Management of Forests in Europe. https://www.foresteurope.org/docs/MC/MC_helsinki_resolutionH1.pdf. Accessed 20 Feb 2021

Intergovernmental Panel on Climate Change (2007) Forestry. Fourth assessment report. Working group III: mitigation of climate change. Cambridge University Press, Cambridge

Jandl R, Ledermann T, Kindermann G, Freudenschuss A, Gschwantner T, Weiss P (2018) Strategies for climate-smart forest management in Austria. Forests 9. https://doi.org/10.3390/f9100592

Kauppi P, Hanewinkel M, Lundmark L, Nabuurs GJ, Peltola H, Trasobares A, Hetemäki L (2018) Climate Smart Forestry in Europe. European Forest Institute. https://efi.int/sites/default/files/files/publicationbank/2018/Climate_Smart_Forestry_in_Europe.pdf

Nabuurs G, Lindner M, Verkerk H, Gunia K, Deda P, Michalak R, Grassi G (2013) First signs of carbon sink saturation in European forest biomass. Nat Clim Chang 3:792–796. https://doi.org/10.1038/nclimate1853

Nabuurs GJ, Delacote P, Ellison D, Hanewinkel M, Lindner M, Nesbit M, Ollikainen M, Savaresi A (2015) A new role for forests and the forest sector in the EU post-2020 climate targets. From Science to Policy 2. European Forest Institute https://doi.org/10.36333/fs02

Nabuurs GJ, Delacote P, Ellison D, Hanewinkel M, Hetemäki L, Lindner M (2017) By 2050 mitigation effects of EU forests could nearly double through European climate smart forestry. Forests 8:484. https://doi.org/10.3390/f8120484

Rockström J, Gaffney O, Rogelj J, Meinshausen M, Nakicenovic N, Schellnhuber HJ (2017) A roadmap for rapid decarbonization. Science 355:1269–1271

United Nations Framework Convention of Climate Change (UNFCCC) (2015) Paris Agreement. UNFCCC, Geneva

Vaara I, Björqvist N, Honkavaara T, Karvonen L, Kiljunen N, Salmi J, Vainio K (2018) Climate smart forestry: results report for the forestry project for mitigating climate change. Metsähallitus Forestry Ltd, http://www.e-julkaisu.fi/metsahallitus/Climate_Smart_Forestry/mobile.html#pid=1. Accessed 10 Jan 2021

Verker PJ, Costanza R, Hetemäki L, Kubiszewski I, Leskinen P, Palahí M, Nabuurs G-J, Potočnik J (2020) Climate-smart forestry: the missing link. For Policy Econ. https://doi.org/10.1016/j.forpol.2020.102164

Yousefpour R, Augustynczik ALD, Reyer C, Lasch P, Suckow F, Hanewinkel M (2018) Realizing mitigation efficiency of European commercial forests by climate smart forestry. Sci Rep 8:345. https://doi.org/10.1038/s41598-017-18778-w

Chapter 10
Climate-Smart Forestry Case Study: Czech Republic

Emil Cienciala

Abstract Forestry in the Czech Republic is facing a historically unprecedented, mostly drought-induced decline in spruce-dominated stands, accompanied by an extensive bark beetle infestation that has spread across most of the country. As a result, the share of sanitary felling has dramatically increased, driving the total harvest to record-high levels in recent years. As a result, current forest management in the country practically resembles a crisis management dealing dominantly with unplanned disturbances. The Czech case shows clearly the essential, non-separable linkage between forest adaptation and mitigation—a simple recognition that, without adaptation, there is no mitigation. It also demonstrates the importance of tailoring the general climate smart forestry approach to regional circumstances. The current priorities of Czech forestry must be to halt forest decline, restore the lost vegetation cover on clearcut soils, and intensify adaptive management in order to create resilient forest ecosystems than can cope better with changing climate and extreme climate events.

Keywords Adaptation · Mitigation · Ecosystem carbon balance · Drought · Bark beetle

10.1 Czech Forestry

Climate-smart forestry (CSF) (Nabuurs et al. 2017) is a proposal aimed at complementing current national strategies for implementing actions under the Paris Agreement (Verkerk et al. 2020). Specifically, CSF advocates for measures to better utilise forestry potential to achieve a stronger climate-change mitigation impact in European countries. This chapter examines how this mitigation concept is applicable to the specific conditions and circumstances of the Czech Republic.

E. Cienciala (✉)
IFER – Institute of Forest Ecosystem Research, Jílové u Prahy and Global Change Research Institute of the Czech Academy of Sciences, Brno, Czech Republic
e-mail: emil.cienciala@ifer.cz

© The Author(s) 2022
L. Hetemäki et al. (eds.), *Forest Bioeconomy and Climate Change*, Managing Forest Ecosystems 42, https://doi.org/10.1007/978-3-030-99206-4_10

Currently (as of 2020), forestry in the Czech Republic is facing a historically unprecedented, mostly drought-induced decline in spruce-dominated stands, accompanied by an extensive bark beetle infestation that has spread across most of the country. As a result, the share of sanitary felling has dramatically increased, driving the total harvest to record-high levels in recent years (Fig. 10.1.). Correspondingly, the share of planned harvest interventions (thinning and final cutting) has declined. Thus, current forest management in the country practically resembles a crisis management dealing dominantly with unplanned disturbances.

Several factors have contributed to the current forest decline in the country, perhaps the most important being: (1) The problematic transformation of Czech forestry following the collapse of the communist regime in the early 1990s. This resulted in, among other things, insufficient personnel in the field and the separation of organisational responsibility and actual forest management. The latter has been driven by an inflexible tender model, which has seriously delayed urgent sanitary interventions in infested (or otherwise damaged) forest stands. (2) Inadequate adaptive forest management and a lack of recognition of the risks associated with the changing climate. Despite relevant targets having been formulated in the second Czech National Forest Programme (Krejzar 2008) and later strategic forestry plans in the country, the implementation of adaptive forest management has been insufficient to significantly increase the resiliency of forest stands. There has been insignificant support for adopting more progressive, close-to-nature forest management, avoiding the clearcut model of even-aged monocultures, utilising natural regeneration, or adequately changing the species and structural composition of forest stands. (3) Objectively exceptional drought conditions and heatwaves in Central Europe in

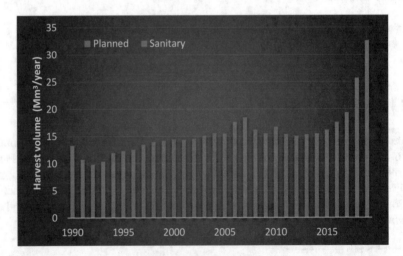

Fig. 10.1 Harvest volume for the period 1990–2019 showing planned and sanitary felling interventions. In 2019, the share of sanitary felling of the total harvest reached 95% at the country level. Apart from the volume of 32.6 Mm³ extracted from the forest in 2019, there is an additional ~6.3 Mm³ of unprocessed wood volume, mostly of dead standing trees, due to an insufficient harvesting capacity. (Czech Statistical Office 2020)

Table 10.1 Selected key characteristics of Czech Forestry

	1990	2018[a]
Forest ownership	95% public	55% public
Employment in forestry	58,600	13,600
Forest area and forestation[b]	2.63 Mha, 33%	2.67 Mha, 34%
Share of conifers[c]	76%	70%
Growing stock per hectare[d,e]	218 m³/ha	270 m³/ha
Harvest level (5-year period)	11.8 Mm³/year (1988–1992)	22.3 Mm³/year (2015–2019)
Share of sanitary felling[f]	47% (1988–1992)	70% (2015–2019)

[a]Data as of 2018, unless stated otherwise
[b]Cadastral forest area, excluding tree vegetation elsewhere
[c]Data from forest management plans linked to the cadastral forest area
[d]Data from forest management plans (sample-based statistical forest inventory data show significantly higher values)
[e]All wood-volume data represent merchantable underbark dimensions (minimum diameter 7 cm)
[f]The relatively high proportion of salvaging around 1990 reflects a period of a significant air-pollution impact on forest stands, manifested by both direct damage to tree foliage and soil disturbed by acidification and nutrient degradation. The latter harmful effect on the soils remains apparent today, with acidified soils impacting root systems and the mycorrhiza, leading to a greater sensitivity of the (mainly coniferous) trees to drought

recent years (Zalud et al. 2020). (4) An inadequate response from the responsible state authorities and the Czech Forests state enterprise to the accelerating forest dieback in the country (Czech News Agency 2020). The Czech Forest Act includes specific expectations for forest owners or entrusted bodies to fulfil their forestry obligations—for various reasons, mostly linked to issue 1, above—and these have not been adequately executed.

In the next section, the CSF principles are outlined, and their alignment with the urgent need to manage current Czech forest decline, create more resilient ecosystems and improve the outlook for Czech forestry (Table 10.1) is assessed.

10.2   Climate-Smart Forestry in the Czech Context

10.2.1   The CSF Concept

The CSF concept (Nabuurs et al. 2017, 2018; Verkerk et al. 2020) builds on three pillars: (1) as a climate-mitigation service, by enhancing carbon storage in forests and wood products, in conjunction with other ecosystem services; (2) through adaptive forest management to increase resilience and improve the health of forest stands; and (3) in the substitution of non-renewable carbon-intensive materials, by using sustainably produced wood resources. The CSF concept represents a more mature strategy than the early, overly carbon-accounting-focused approaches and policies, such as those driven by the Kyoto Protocol. Those policies prioritising mitigation actions using forest resources disregarded the following essential aspects (and not only from the Czech forestry point of view): (1) the long-term forestry production cycle; (2) the importance of other ecosystem services, such as water retention, soil protection and biodiversity; and (3) perhaps most importantly, the essential, non-separable linkage between forest adaptation and mitigation—a simple recognition that, without adaptation, there is no mitigation. In other words, specifically under changing environmental conditions, adaptation management must be prioritised in order to secure the sustained provisioning of ecosystem services, including climate mitigation. Failure to adequately adapt forests and forestry (within an appropriate time and scope) is predestined to result in undesired reverse effects, with forestry turning into a significant source of emissions instead of the expected sink. This risk is being increasingly internationally recognised (Anderegg et al. 2020), and is also plainly demonstrated by the current situation in Czech forestry, as detailed in the following sections.

CSF clearly links essential ecosystem services, stressing adaptation to secure forest health and increase resilience, and promoting the important substitution function that wood products offer. However, CSF should also consider other fundamental constraints, such as governance issues and the legacy of past management. When reviewing the factors responsible for the current forest decline in the Czech Republic, as highlighted above, it becomes clear that a holistic CSF approach also needs to address factors relating to governance, business models and/or specific management actions against bark beetle outbreaks (Hlasny et al. 2019).

10.2.2   Forestry-Based Climate Mitigation

An earlier CSF case study using the Czech Republic (Nabuurs et al. 2018) included, among other things, a mitigation-impact projection based on a calibration period up to 2015. The model projection up to 2100 reported that the anticipated adaptive management would result in a smaller sink in Czech forests in relation to the business-as-usual scenario, only providing additional mitigation benefits after 2080, together with more resilient forest stands. That study, however, did not anticipate the scale of the current drought-induced forest decline that the Czech Republic has been experiencing since 2015. This development significantly and negatively affects the mitigation outlook for Czech forestry for the coming decades.

The carbon budget of Czech forestry, as reported in the recent greenhouse gas emissions inventory submission (National Inventory Report, CHMI 2020) is illustrated in Fig. 10.2. The effect of the recent decline in coniferous forest stands is obvious in the rapidly declining sink that became an emissions source in 2018, for the first time since 1990. This means that the forest sector, which used to offset about 6% of Czech national emissions, has turned into yet another source category, with a notable magnitude of emissions. The contribution of harvested-wood products (HWPs) still counts, mostly acting as a sink in the Czech circumstance, and corresponding to the generally increasing total harvest volume (Fig. 10.1). However, the annual offset represented by HWPs is estimated to be about 1 Mt. CO_2 for the

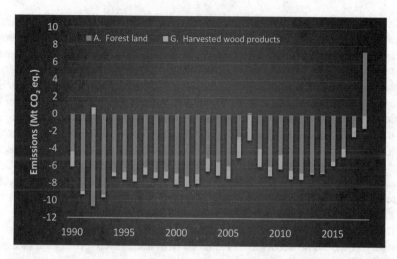

Fig. 10.2 Emissions contribution from the Czech forestry sector (preliminary data of IFER as of April 2020). Negative values represent the sinking of emissions, positive values represent a source of emissions. The data shown distinguish between the major United Nations Framework Convention on Climate Change emissions categories: A—forest land, where emissions are determined mainly based on changes in living biomass; G—HWPs, where emissions are determined based on changes in this pool

period 1990–2020, which is small in comparison to the total emissions of the country (134 Mt. CO_2 eq. as of 2018). Overall, the emissions contribution from forestry, including the HWPs offset, is currently 4.5% of the total emissions of the country (Czech Hydrometeorological Institute 2020).

Data on woody material used for bioenergy remains uncertain, but has been estimated to represent about 12% of the total harvest, annually (i.e. close to 2 Mm^3 of wood volume or about 1 Mt. of biomass), as of 2018. However, the extent to which this amount contributes indirectly as a substitution effect in the energy sector is difficult to ascertain due to the inherent uncertainty in the related statistics and/or missing information.

In addition, there is the contribution of fast-growing woody plantations for energy purposes, which is commonly accounted for under agriculture. In the Czech Republic, the spatial extent of such systems is only about 3 kha, and this has stagnated for various reasons, with the result of currently producing only a marginal climate-mitigation effect.

10.2.3   Towards Adaptive Forest Management

The forest policy-makers in the country did recognise the need to significantly improve forest-stand resilience, specifically in the second National Forestry Programme (Krejzar 2008). In the progamme's Key Action 6, 12 measures were defined and developed, their implementation designed to alleviate the impact of climate change and extreme events on the forest sector (Cienciala 2012). It stressed the need to grow diversified forest stands, employing the maximum use of natural processes, diverse species compositions, natural regeneration, and a spectrum of silvicultural practices to enhance the resiliency of forest stands. The accompanying measures included specific actions to broadly support the main goals. Evidently, implementation of these measures has been too slow and has not gone far enough to reduce the large-scale, drought-induced decline and bark beetle outbreak in spruce-dominated coniferous stands (as well as in pine and larch), as witnessed in recent years (Hlasny et al. 2019). For example, the spatial representation of more-resilient broadleaved tree species has increased by only by 5% in 2000–2018. Similarly, the share of natural regeneration has only increased from 13.5 to 16.1% in the same period (Ministry of Agriculture 2019).

The more-recent National Action Plan on Adaptation in the Czech Republic (Ministry of the Environment 2017) stressed two fundamental prioritized measures applicable for the forestry sector, namely

1. Support of the natural adaptive capacity of forests and strengthening of their functioning under changing climate; and
2. protection and revitalisation of the natural water regime in forests.

The explicit implementation issues in the above measures have been undermined by a sustained preference for the clearcut system and linked forestry operations, and the unsupportable hoofed game stocks that effectively hinder use of natural regeneration.

Obviously, these measures implicitly recognise that the precondition for any mitigation effect realised in Czech forestry is to ensure the resiliency of forest stands under changing climate conditions, with a specific focus on the water regime and the prevention of drought. Also important to note is the emphasis on soil conditions—the elementary resource for life and an essential part of forest ecosystems. Only functional forest ecosystems can deliver the spectrum of expected ecosystem services, with climate mitigation being only one of these, and being fully dependent on the success of the adaptation measures.

The evolution of the situation in Czech forestry has led to the swift adoption of an actual guiding forest-policy document—Conception of the governmental forestry policy until 2035 (Ministry of Agriculture 2020). Its declared four long-term goals are:

1. Ensure sustained and full provisioning of all of forest ecosystem services for future generations.
2. With respect to changing climate, increase biodiversity and the ecological stability of forest ecosystems while retaining their productive functions.
3. Ensure competitiveness in forestry and linked sectors, and their importance in regional development.
4. Enhance advisory services, education, research and innovation in forestry.

The development of Czech forest policy reflects the growing urgency and better comprehension of the wide role of forest resources in society, with a notably increased accentuation of environmental services and sustainability. In this respect, Czech forest policy is becoming fully consistent with the current trends in European policy, as expressed by Forest Europe, among other entities. For example, the Concept (of forestry policy: Ministry of Agriculture 2020), in its goals, accords with the recently announced ambitions of the European Commission's Mission on Soil Health and Food (Veerman et al. 2020) under the Horizon Europe Research and Innovation Programme, which also concerns forestry.

The CSF principles align, in part, with the declared long-term goals of the Czech Concept up to 2035. Specifically, the CSF pillars 2 (adaptive forest management) and 3 (enhanced use of HWPs) address the goals of Concept 2035. However, it is still to be seen to what extent the CSF's primary goals of enhanced carbon storage mitigation is prioritised as part of the current Czech forestry strategy, which has very much been focused on the immediate management of the current local environmental crisis.

10.2.4   Mitigation Outlook for Czech Forestry

What is expected, in terms of mitigation, from the Czech forestry sector in the coming decades? The most recent outlook for mitigation was presented in the Czech National Forest Accounting Plan (Ministry of the Environment 2017), in conjunction with setting a national forest reference level (FRL) under EU regulation 2018/841. The two presented scenarios were prepared using the calibrated CBM-CSF3 model (Kull et al. 2016), and present a rather pessimistic outlook for Czech forestry, in which it is projected to lose much of its carbon sequestration capacity in the coming decades. In Fig. 10.3, we show two corresponding scenarios of emissions (red and black), together with a third scenario (green) representing the most up-to-date (as of June 2020) projection estimates (IFER—unpublished data 2019) for combatting the current drought-induced bark beetle outbreak.

Each of these scenarios include emissions from the change in biomass carbon stock and the HWP contribution, combined. The red scenario represents a development with re-occurring bark beetle outbreaks each decade, whilst the black scenario shows the pessimistic outlook of bark beetle outbreaks resulting in a reduction in spruce growing stock by 80% by 2050, with the corresponding remaining areal representation of spruce reduced to 10–15%, compared to ~50% as of 2018. The green scenario counts on a more rapid stabilisation of forest health, with a resulting reduction in spruce management approaching 20% by forest area. This green scenario means a return to an overall carbon sink in the forest (biomass + HWPs) by 2030. The sink capacity would then remain strong for the following two decades, mainly due to the significantly reduced harvesting potential of conifers and the only gradually increasing harvesting possibilities in broadleaved tree species.

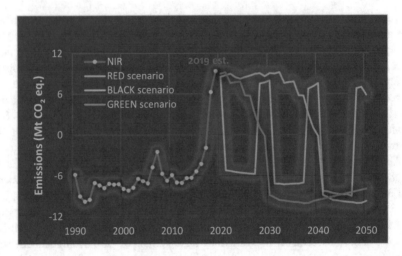

Fig. 10.3 Outlook of the emissions balance in Czech forestry to 2050, expressed by three scenarios, each including the biomass carbon pool and the contribution of harvested wood products (IFER—unpublished data as of June 2020). See the text for the narrative of individual scenarios

Obviously, this possible development will have a significant impact on the economics of forest owners. The requisite sanitary measures are, and will remain, costly, and will not be compensated for by wood sales. As is already known, the wood price has dropped significantly due to a current wood oversupply in European countries. The Czech Republic currently sells its wood also to China, despite low prices and increased transaction costs. The fact that the country is not able to increase its own wood-processing capacity to make products with increased value does not help the situation. However, with respect to the key mitigation instrument of the EU—regulation 841/2018 on accounting for the land use, land-use change and forestry sector during the 'Paris' period of 2021–2030—the Czech Republic faces a challenge. The key element of this regulation is setting the FRL based on management practices (including harvesting level), as of 2000–2009. With the current down-correction imposed by the European Commission, the FRL for the Czech Republic is set to ~6.1 Mt. CO_2 eq. for 2021–2025. With respect to the projected emissions in the green scenario, this FRL would be surpassed by 14 Mt. CO_2 annually, representing an unforeseen economic loss for the country, realised through the emissions allowance system. Clearly, there is a need to adjust the reference level using technical correction, considering the major changes Czech forests have been experiencing recently as a result of the disturbances.

10.3   Conclusions and Policy Implications

The case of the Czech Republic shows the importance of tailoring the general CSF approach to regional circumstances. The bark beetle disturbance/calamity of recent years has progressed to such a massive scale that the most immediate task for CSF would be to enhance the adaptation and resilience of Czech forests, thereby also seeking to increase their carbon sink potential in the long term. One important part of CSF in the Czech case is that it needs to be institutional, organisational and have governance aspects in general. These need to be fine-tuned and updated in a way such that more efficient measures can be undertaken for adapting and improving forest resilience, and sustaining the long-term environmental and social requirements of Czech forests. It is further stressed that:

1. Mitigation is to be understood as only one of the vital ecosystem services provided by forest ecosystems. Under the Czech conditions, soil protection, water regulation and hosting biodiversity are increasingly being recognised as priority services that are important to local society.
2. Preserving and increasing carbon storage in the long term is possible only through the establishment of healthy, resilient and sustainably used forest ecosystems, well adapted to changing growth environments, which implies that adaptation must be prioritised to ensure a sustained mitigation effect from forestry.

The current priorities of Czech forestry must be to halt forest decline, restore the lost vegetation cover on clearcut soils, and intensify adaptive management in order to create resilient forest ecosystems than can cope better with changing climate and extreme climate events.

References

Anderegg WRL, Trugman AT, Badgley G, Anderson CM, Bartuska A, Ciais P, Cullenward D, Field CB, Freeman J, Goetz SJ, Hicke JA, Huntzinger D, Jackson RB, Nickerson J, Pacala S, Randerson JT (2020) Climate-driven risks to the climate mitigation potential of forests. Science 368(6497):eaaz7005. https://doi.org/10.1126/science.aaz7005
Cienciala E (2012) Key Action 6 – Mitigate impacts of expected global climate change and extreme meteorological events. Lesnicka práce 4, 232–234, in Czech
Czech Hydrometeorological Institute (CHMI) (2020) National greenhouse gas inventory report, NIR (reported inventory 2018). CHMI, Prague. https://unfccc.int/ghg-inventories-annex-i-parties/2020. Accessed 16 June 2021
Czech News Agency (CNA) (2020) Too slow reaction of Czech Forest, S.A., on the bark beetle outbreak, stated the Czech Republic Supreme Audit Office. https://zpravy.aktualne.cz/domaci/na-kurovcovou-kalamitu-reagovaly-lesy-cr-pomalu-zjistil-kont/r~6734c80ca94f11eab115ac1f6b220ee8. Accessed 16 June 2021, in Czech
Czech Statistical Office (2020) Forestry - 2019. The publication Forestry gives overall information on forestry activities in the Czech Republic, accessible digitally at https://www.czso.cz/csu/czso/lesnictvi-2019

Hlásny T, Krokene P, Liebhold A, Montagné-Huck C, Müller J, Qin H, Raffa K, Schelhaas M-J, Seidl R, Svoboda M, Viiri H (2019) Living with bark beetles: impacts, outlook and management options. From science to policy 8. European Forest Institute, Joensuu

Krejzar T (ed) (2008) National Forest Programme for the period until 2013. On the authority of the Ministry of Agriculture. Forest Management Institute, Prague

Kull SJ, Rampley GJ, Morken S, Metsaranta J, Neilson ET, Kurz WA (2016) Operational-scale carbon budget model of the Canadian forest sector (CBM-CFS3): version 1.2, user's guide. Northern Forestry Centre, Canada

Ministry of Agriculture (MA) (2019) Report on the state of forests and forestry in the Czech Republic in 2018. MA, Prague (in Czech)

Ministry of Agriculture (MA) (2020) Conception of the government forestry policy until 2035. MA, Prague (in Czech)

Ministry of the Environment (ME) (2017) National action plan on adaptation to climate change. ME, Prague (in Czech)

Nabuurs GJ, Delacote P, Ellison D, Hanewinkel M, Hetemäki L, Lindner M (2017) By 2050 the mitigation effects of EU forests could nearly double through climate smart forestry. Forests 8:1–14. https://doi.org/10.3390/f8120484

Nabuurs G-J, Verkerk PJ, Schelhaas M-J, González Olabarria JR, Trasobares A, Cienciala E (2018) Climate-smart forestry: mitigation impacts in three European regions. European Forest Institute, Joensuu

Veerman C, Correia TP, Bastioli C, Biro B, Bouma J, Cienciala E, Emmett B, Frison EA, Grand A, Hristov L, Kriaučiūnienė Z, Pogrzeba M, Soussana J-F, Vela C, Wittkowski R (2020) Caring for soil is caring for life. European Commission, Brussels. https://op.europa.eu/en/publication-detail/-/publication/4ebd2586-fc85-11ea-b44f-01aa75ed71a1. Accessed 21 Jan 2021

Verkerk PJ, Costanza R, Hetemäki L, Kubiszewski I, Leskinen P, Nabuurs GJ, Potočnik J, Palahí M (2020) Climate-smart forestry: the missing link. For Policy Econ 115:102164. https://doi.org/10.1016/j.forpol.2020.102164

Zalud Z, Trnka M, Hlavinka P et al (2020) Agricultural drought in the Czech Republic – development, impact and adaptation. Agricultural Chamber of the Czech Republic, Prague (in Czech)

Chapter 11
Climate-Smart Forestry Case Study: Finland

Heli Peltola, Tero Heinonen, Jyrki Kangas, Ari Venäläinen, Jyri Seppälä, and Lauri Hetemäki

Abstract Finland is the most forested country in the EU – forests cover 74–86% of the land area, depending on the definition and source. Increasing carbon sequestration from the atmosphere, and by storing it in forests (trees and soil) will be one important part of the Finnish climate smart forestry strategy. However, just maximizing the carbon storage of forests may not be the best option in the long run, although it may provide the best climate-cooling benefits in the short term. This is because the increasing risks of large-scale natural disturbances may turn forests, at least partially, into carbon sources. The climate change adaptation and mitigation should therefore be considered simultaneously. Different adaptation and risk management actions will be needed in Finnish forests in the coming decades to increase forest resilience to multiple damage risks. This could be done, for example, by increasing the share of mixtures of conifers and broadleaves forests instead of monocultures. Yet, the CSF strategy should also include the production of wood-based products that act as long-term carbon storage and/or substitute for more GHG-emission-intensive materials and energy. Doing this in a way which also enhances biodiversity and sustainable provisioning of multiple ecosystem services, is a key. Moreover, increasing forest land – for example, by planting on abandoned or low-productivity agricultural land, especially on soils with a high peat content – would enhance climate change mitigation.

Keywords Forestry · Boreal forests · Climate change · Adaptation · Risk Management · Climate change mitigation ·

H. Peltola (✉) · T. Heinonen · J. Kangas
University of Eastern Finland, Joensuu, Finland
e-mail: heli.peltola@uef.fi

A. Venäläinen
Finnish Meteorological Institute, Helsinki, Finland

J. Seppälä
Finnish Environment Institute, Helsinki, Finland

L. Hetemäki
European Forest Institute, Joensuu, Finland

Faculty of Agriculture and Forestry, University of Helsinki, Helsinki, Finland

© The Author(s) 2022
L. Hetemäki et al. (eds.), *Forest Bioeconomy and Climate Change*, Managing Forest Ecosystems 42, https://doi.org/10.1007/978-3-030-99206-4_11

11.1 Finland's Forest Resources and Their Utilization

Finland is the most forested country in the EU, in terms of land area (Table 11.1). Depending on the definition of forest land, and the source, forests cover 74–86% of the land area. Finland's forests account for around 14% of the total EU27 forest land. The volume of growing stock and increments have almost doubled in the past five decades (Fig. 11.1; Finnish Forest Statistics 2019, 2020). Improved forest

Table 11.1 Overview of Finnish forest sector

Forest resources	
Area of forest land	26.3 million ha, of which 77% is productive and 10% poorly productive (the rest is unproductive land, forest roads, etc.)
Strictly protected forest area and biodiversity conservation areas in commercial forests	2.2 and 0.5 million ha
Total volume of growing stock	2482 Mm^3
Carbon storage of forest land	3200 Mt. CO_2-eq. in forest biomass and 14,000 Mt. CO_2-eq. in soil (most soil carbon in peatlands)
Net carbon sink of forest land	25.6 Mt CO_2-eq. in forest land in 2019, corresponding to 48% of total GHG emissions in Finland (additionally, 3.4 Mt CO_2-eq. in wood-based products, with estimated substitution impact of 27 Mt CO_2-eq.)
Average annual growth of growing stock	108 Mm^3 year^{-1}
Total volume of harvested roundwood	≈78 Mm^3 in 2018, this year being the all-time high
Total drain (harvested roundwood, logging residues and natural drain)	≈ 94 Mm^3 in 2018
Average growing stock volume on forest land (productive/ poorly productive land)	119 m^3 ha^{-1}
Tree species composition	50% scots pine, 30% Norway spruce, 17% silver and downy birch, and 3% other broadleaves
Ownership	
Private	52%
State	35%
Companies	7%
Municipalities, parishes, funds, associations	6%
Economic contribution	
The value added in the forest sector	9 billion euros in 2019, 4.3% of the national economy
Employees in the Finnish forest sector	66,000 (forestry 26,000, forest industries 40,000) in 2019

Source: Finnish Forest Statistics (2019, 2020), Statistics Finland (2020)

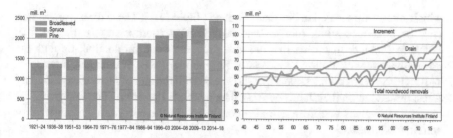

Fig. 11.1 Development of growing stock volume on forest land and poorly productive forest land, and total roundwood removals, increment and drain of the growing stock, in past decades in Finnish forests. (Sources: Finnish Forest Statistics 2019, 2020)

management practices have largely contributed to this change (Finnish Forest Statistics 2020). The forest growth has been increased through the ditching of peat-lands, forest fertilization, maintaining higher growing stock (per hectare) in fre-quent thinnings, regeneration of poorly productive forests, and using improved forest regeneration methods and materials (seedlings and seeds), respectively. Additionally environmental change (e.g. climate change and nitrogen deposition) has contributed to this change (Henttonen et al. 2017). Another reason for this change is that annual wood removal in the last five decades, has been, on average, clearly less than the increment of the forests.

On the other hand, intensified forest management targeting for increased wood production has also affected harmfully forest biodiversity and the provisioning of some ecosystem services (Lehtonen et al. 2021). Also, the use of forest fertilization, and ditch network maintenance in peatland forests, have increased nutrient leaching and carbon emissions from the soil (Finér et al. 2020; Lehtonen et al. 2021). Until recently, the management of Finnish forests has been based, almost solely, on even-aged rotation forestry. However, interest among forest owners, professionals and general society in diversifying forest management practices and increasing provi-sioning of multiple ecosystem services has increased the attractiveness of uneven-aged management and mixed-species forestry (Díaz-Yáñez et al. 2020).

According to the National Resources Institute Finland, the maximum sustainable roundwood removal potential of Finnish forests, on land assigned for timber pro-duction, is 84 Mm^3 year^{-1}, on average, for 2015–2024. Annual wood removal in recent years has corresponded to an average of 75% of the total forest growth, which includes the growth of strictly protected forests and natural drainage (Fig. 11.1). This percentage is clearly higher for Finnish forests compared to the EU average, with Finland's forest sector having a relatively bigger role in the country's economy than is the case for any other EU country. Altogether, around 620,000 private forest owners sell about 80% of the Finnish forest industries' total domestic wood supply. Thus, the income generated by forestry is spread among a relatively large part of the population.

Besides the demand for wood production, there are increasingly high demands for forest-related recreation, tourism, biodiversity and forest carbon sinks. No for-estry measures are allowed on 10% of the most strictly protected forested areas, which are mostly located in Northern Finland (Table 11.1). These forests are important for both recreation, tourism and biodiversity. Biodiversity is preserved in Southern Finland, in forests that are also used for wood production, through the government-funded Forest Biodiversity Programme (METSO, annual funding of 7–10 million euros). This targets forest owners, with the aim of increasing voluntary forest protection on their lands by 96,000 ha by 2025. Preserving and improving the biodiversity values of forests are also considered in the everyday management of commercial forests.

There are increasing EU and domestic pressures to increase the capacity of the forest carbon sink in Finland, such as the Green Deal and the updating of the land use, land-use change and forestry (LULUCF) regulation. According to the LULUCF regulation, for Finland, the reference level for a forest carbon sink with forest products is 29.4 million CO_2- equivalent (CO_2-eq.) tons in 2021–2025 (Suomen ilmastopaneeli 2021). However, the actual forest carbon sink can vary significantly from one year to the next, along with the annual harvesting levels (since the 1990s, these have been 17.5–47.4 Mt CO_2-eq., annually), which are largely affected by forest-industry business cycles. For example, in 2018 the forest carbon sink was clearly lower than in the previous and succeeding years, due to a higher total annual volume of harvested roundwood (Table 11.1). Given the various demands on Finnish forests, it is necessary to find a balance. Moreover, it is crucial to try to minimize the trade-offs and maximize the synergies between the different uses of forests. In this case-study, we analysed what the climate-smart forestry (CSF) approach could mean in the Finnish context in the coming decades.

11.2 Impacts of Changing Climate, Forest Management and Harvesting

11.2.1 Development of Forest Resources and Carbon Sinks

Compared to the reference period of 1981–2010, the annual mean temperature in Finland may increase by 1.9–5.6 °C and the mean annual precipitation by 6–18% by the 2080s under different GHG scenarios (i.e. Representative Concentration Pathways, RCPs) (Ruosteenoja et al. 2016). Forest growth is generally projected to increase significantly more in the northern boreal zone of Finland than in the southern boreal zone (Fig. 11.2), due to the differences in prevailing climatic conditions (e.g. temperature, precipitation) and forest structure (e.g. age and tree species composition) in these regions (Kellomäki et al. 2008). Overall, the increase in forest growth will come from birch (*Betula* spp.), in particular, but also Scots pine (*Pinus sylvestris*) (Kellomäki et al. 2018). For Norway spruce (*Picea abies*), the growing conditions may become suboptimal, especially in the southern boreal zone, along with increasing summer temperatures and drought.

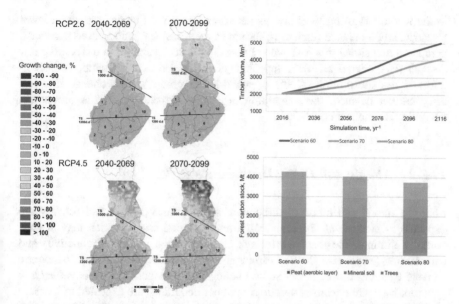

Fig. 11.2 Left: Spatial distribution of the percentage change in tree growth (diameter) in Finland over all tree species on upland (mineral) forest inventory plots, given separately for the coming two 30-year periods (2040–2069 and 2070–2099), under the RCP2.6 and RCP4.5 scenarios, compared to the period 1981–2010 (Kellomäki et al. 2018). If considering peatlands, the positive and negative impacts would be slightly stronger. The temperature sum lines across the country separate the southern (TS > 1200 d.d.), central (1000 d.d. < TS < 1200 d.d.), and northern (TS < 1000 d.d.) boreal regions. Right: Timber volume development (top) and average carbon stock in trees and soil (bottom) on forest land currently available for timber production in Finland in 2016–2116, with scenarios of 60-, 70- and 80-Mm³ year⁻¹ timber cutting targets under the RCP2.6 scenario, with intensified forest management (data from Seppälä et al. 2019). The increasing abiotic and biotic damage risks under climate change were not considered in these scenario analyses

In addition to the severity of climate change, the intensity of forest management and harvesting will also affect the future development of Finnish forests, and consequently timber supply, the carbon sink and the balance of forestry (e.g. Hynynen et al. 2015; Heinonen et al. 2017, 2018). If assuming mild (RCP2.6) climate change and annual mean timber harvests of 60–80 Mm³ year⁻¹, the average annual volume increment could be increased by 4.5–5.7 Mm³ year⁻¹ in 2016–2116, and timber volume may reach 2.7–5.0 Bm³ by 2116, on forest land currently available for timber production (Fig. 11.2), if increasing the use of forest fertilization and improved regeneration material (Heinonen et al. 2018).

Forest biomass contributes about 23–30% to the total carbon stock of forests (in trees and soil, including mineral soil and the aerobic layer of peat) (Fig. 11.2). Maintaining lower harvesting levels increases the carbon sink and the balance of forests (in trees and soil), but it decreases the carbon stock in wood-based products. Overall, the long-term carbon stock of wood-based products is small compared to that of forest biomass and soil. This is because a relatively small share of harvested wood is used in wood-based products with long-life cycles. In this sense, an increase in the wood harvesting level always results in less carbon being sink and a lower forestry carbon balance (including carbon in the forest and wood-based products), compared to a situation where wood harvesting is not increased in the coming few decades (Heinonen et al. 2017; Seppälä et al. 2019). On the other hand, a consideration of the substitution effects of wood-based products may change this forestry carbon balance, the magnitude of change depending on the production portfolio (Hurmekoski et al. 2020).

11.2.2 Abiotic and Biotic Disturbance Risks

Multiple abiotic and biotic disturbance risks to Finnish forests and forestry are expected to increase at different spatial and temporal scales, which may at least partially eliminate the positive effects of climate change on forest productivity and carbon sinks (Reyer et al. 2017). Warmer and wetter winters are expected to increase damage by windstorms, heavy snow loading and pathogens (e.g. *Heterobasidion* spp, root rot), while warmer and drier summer conditions are expected to increase insect pests (e.g. European spruce bark beetle, *Ips typographus*), droughts and forest fires, particularly in coniferous forests. The occurrence of different damaging agents (excluding snow extremes) is expected to increase, especially in southern and middle Finland (Mäkelä et al. 2014; Lehtonen et al. 2016a, b; Ruosteenoja et al. 2018; Venäläinen et al. 2020). A shortening of the soil frost period from late autumn to early spring will increase the wind damage risk, despite no great change in the wind regime (Lehtonen et al. 2019). Wind- and snow-damaged timber left in the forest will increase the amount of breeding material for bark beetles, an outbreak of which may, together with drought, further increase forest fire risk, through increased amounts of easily flammable deadwood. Attacks by *Heterobasidion* species may increase due to increasing tree injuries during harvesting in the unfrozen soil season (Honkaniemi et al. 2017). Wood decay will also increase the risk of wind damage due to poorer anchorage and stem resistance of trees.

11.3 Nexus for Adaptation, Resilience and Mitigation of Climate Change

11.3.1 Adaption to Climate Change and Risk Management

Different adaptation and risk management actions will be needed in Finnish forests in the coming decades in order to adapt appropriately to climate change and to increase forest resilience to multiple damage risks (Venäläinen et al. 2020). Possible adaptation and risk management actions evaluated in Finland, have so far considered almost solely even-aged forestry. However, some of these are also applicable to uneven-aged and mixed-species forestry.

In the southern boreal zone, a decrease in the cultivation of Norway spruce may be needed, particularly on forest sites with a relatively low water holding capacity, which are more suitable for Scots pine. Also, the potential for an increase in spring and summer droughts should be considered when planting seedlings or seeding in order to increase the success of forest regeneration. Additionally, by favouring growing mixtures of conifers and broadleaves (e.g. spruce and pine, spruce and birch, or pine and birch) instead of monocultures, forest resilience may be increased against multiple damage risks. Overall, timely precommercial thinning and more frequent or heavier commercial thinnings may also be needed in order to increase forest resilience and forest growth and to avoid an increase in natural mortality in stands that are too dense. A shortening of the rotation length may also be needed in order to increase forest resilience, especially for Norway spruce, which may be subject to multiple forest disturbance risks (e.g. wind damage, drought, European spruce bark beetle and *Heterobasidion* spp, root rot).

In planning and implementing thinnings and clearcuts, the increasing risks of wind damage should be considered, especially in the southern and central boreal zones, where strong winds will blow more frequently under unfrozen soil conditions (Laapas et al. 2019). Especially on high-risk areas, heavy thinnings should be avoided on the upwind edges of new clear cuts, and the creation of large height differences should be avoided between adjacent stands in the final harvesting, respectively (Heinonen et al. 2009). It is also recommended that forest fertilization is avoided at the same time as thinning in high-risk areas for wind and snow damage. Consequently, in the middle and northern boreal zones, timely precommercial and commercial thinning may increase the resilience of Scots pine and birch stands to snow damage. Also, the avoidance of forest fertilization on forest sites at high altitudes is suggested in order to decrease snow damage risks, regardless of tree species (> 200 m above sea-level). Timber damaged by wind and snow should also be harvested in a timely manner and transported out of the forest (also undamaged harvested timber) in order to avoid unnecessarily increasing the amount of breeding material for bark beetles. This also holds for bark-beetle-infested and *Heterobasidion*-infected trees.

Because climate change will induce multiple damage risks in Finnish forests and forestry, the probability of devastating cascading events is also projected to increase (Venäläinen et al. 2020). However, their severity may vary significantly at different spatial and temporal scales. Therefore, frequent adjustments to forest management practices in response to changing growing conditions will be required, in order to adapt to climate change and maintain forest resilience. This is also important from the climate change mitigation point of view because large-scale natural disturbances may act as significant carbon sources (Kauppi et al. 2018). Therefore, the multiple risks to forests need to be considered simultaneously in the planning and implementation of forest management. The flexible use of diverse management strategies, instead of one single management strategy (e.g. even-, uneven- and any-aged management) may help to ensure forest resilience and simultaneously provide multiple ecosystem services for society (Díaz-Yáñez et al. 2020).

11.3.2   Climate Change Mitigation

A forest-rich country like Finland can contribute to climate change mitigation especially by increasing carbon sequestration from the atmosphere, and by storing it in
forests (trees and soil), but also by producing wood-based products that can act as
long-term carbon storage and/or can substitute for more GHG-emission-intensive
materials and energy (Hurmekoski et al. 2020). Whether the carbon sink of Finnish
forests (and the forest sector) will remain at the current level or increase/decrease in
the future will strongly depend on the intensity of forest management and harvesting related to wood demand in the coming decades (see Heinonen et al. 2017, 2018).
The carbon sink will also be affected by the severity of climate change and natural
disturbances (Venäläinen et al. 2020).

In order to increase the climate benefits of harvested wood, it should be increasingly used for products and fuels that will release fewer GHG emissions to the
atmosphere than the fossil-based products and fuels it is substituted for (Hurmekoski
et al. 2020). However, the substitution effects must be, on average, even doubled for
additional wood harvest if, for example, 80 Mm^3 year^{-1} is harvested annually instead
of 60 Mm^3 year^{-1} (in the coming 100 years) (Seppälä et al. 2019). This would be
needed in order to compensate the lower carbon stocks of Finnish forests with
increased harvest levels (Seppälä et al. 2019). On the other hand, lower harvesting
levels in Finland would most likely increase harvesting in other countries. In the
longer term, all sustainable uses of renewable wood that compensate for the use of
fossil resources might be seen as remaining beneficial because we should be giving
up using fossil resources as soon as possible, from the viewpoint of mitigating climate warming in the long term.

Forest growth will also decline, along with aging, which, together with a large
volume of growing stock, could promote multiple natural disturbances and, consequently, carbon release into the atmosphere over the long term. Old-growth forests
also sequestrate less carbon than younger forests, but they may offer significant
carbon storage (Gundersen et al. 2021; Kellomäki et al. 2021). Forests also contribute to several other climate impacts, in addition to GHG emissions (e.g. albedo,
biogenic aerosols, evaporation and surface roughness), which may be affected,
directly or indirectly, through changes in forest cover and structure, and by the
intensity of forest management and harvesting (Kalliokoski et al. 2020; Kellomäki
et al. 2021). The opposing effects of changes in albedo and carbon stocks may also
largely cancel each other in managed forests with little remaining net climate effect
(Kellomäki et al. 2021). In short term, no management option may provide larger
net climate benefits than even-aged or uneven-aged management, but increasing use
of this option may require proper incentives such as compensation for lost harvest
incomes for forest owners.

11.4   Climate-Smart Forestry Strategies and Policy Measures

Despite the important role of the Finnish forest sector in the national GHG balance, maximizing the carbon storage of forests may not be the best option in the long run, although it may provide the best climate-cooling benefits in the short term. This is because an increase in large-scale natural disturbances (e.g. storms, forest fires and European spruce bark beetle outbreaks) may turn forests, at least partially, into carbon sources that release large amounts of carbon into the atmosphere. Instead, in CSF, it is preferential to both increase the carbon stocks and sinks in forests, and increasingly use harvested wood for products and fuels, which will release fewer GHG emissions into the atmosphere, rather than the fossil-based products and fuels they are substituting for. At the same time, maintaining biodiversity and sustainably provisioning multiple ecosystem services should be ensured (Heinonen et al. 2017; Díaz-Yáñez et al. 2020).

Overall, living forest biomass and mineral soils (decaying organic matter and soil organic matter) remove carbon from the atmosphere (net carbon sink), and organic soils (peatlands) emit carbon (net carbon source) (Fig. 11.3). The harvesting level affects forest carbon storage and sinks more than forest management practices and ongoing climate change (Heinonen et al. 2017, 2018; Seppälä et al. 2019).

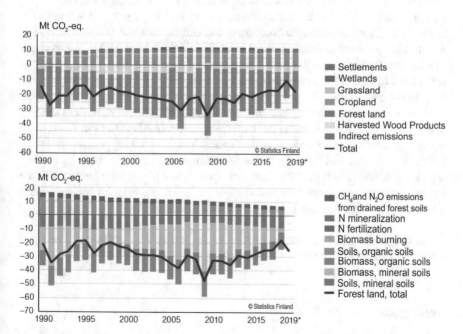

Fig. 11.3 Emissions (positive sign) and removals (negative sign) of Mt CO_2-eq. from different land-use categories (top) and forest land (bottom) in 1990–2019 in Finland (*partial estimation for 2019). (Source: Statistics Finland 2020)

However, forest carbon sinks and storage could be increased in even-aged forestry by increasing the use of improved forest regeneration material and forest fertilization, and by maintaining sufficient growing stock in thinnings (Lehtonen et al. 2021). Also, nutrient leaching and GHG emissions may be decreased on peatlands by maintaining a high enough soil water table level. This could be done by using uneven-aged forestry (especially selective cuttings) on suitable sites, and by avoiding unnecessary ditch network maintenance (Ojanen et al. 2019; Leppä et al. 2020a, b; Finér et al. 2020). This is necessary because a low soil water table will increase CO_2 emissions (Ojanen et al. 2019), and N_2O emissions, especially on fertile peatland sites (Minkkinen et al. 2020). On the other hand, CH_4 emissions may be notable on peatland sites with a high soil water table.

Increasing forest land – for example, by planting on abandoned or low-productivity agricultural land, especially on soils with a high peat content – would be a positive action when it comes to climate change mitigation. On the other hand, also decreasing the deforestation may be effective; currently, deforestation is occurring at a rate of about 10,000 ha, or 0.04% of the total forest area, annually (Kärkkäinen et al. 2019). Increasing the use of by-products for textiles and wood–plastic composites, in place of kraft pulp and biofuel, may also help to provide greater overall substitution credits compared to increasing the level of wood use for construction (Hurmekoski et al. 2020).

To conclude, forests and forest-based bioeconomy can contribute considerably to climate change mitigation in forested countries like Finland, through reducing GHGs in the atmosphere, especially by increasing the carbon sequestration and storage in forests, but also through carbon storage in wood-based products with long life-cycles and the substitution of fossil-intensive resources (Hurmekoski et al. 2020). However, at the same time, there is a pressure to both diversify forest management and increase the provisioning of versatile ecosystem services for society. The forest management implemented today strongly affects the future supply of different ecosystem services (Heinonen et al. 2017). Overall, CSF requires appropriate adaptations of forest management and utilization of forests under climate change, by taking account the multiple risks to forests and forestry. Different management strategies may be needed, depending on the region (and site) and time span, in order to ensure forest resilience and the simultaneous provisioning of multiple ecosystem services for society. This is important also from the climate change mitigation point of view.

Acknowledgments Heli Peltola thanks the financial support from the OPTIMAM project (grant number 317741) funded by the Academy of Finland.

References

Díaz-Yáñez O, Pukkala T, Packalen P, Peltola H (2020) Multifunctional comparison of different management strategies in boreal forests. For Int J For Res 93(1):84–95. https://doi.org/10.1093/forestry/cpz053

Finér L, Lepistö A, Karlsson K, Räike A, Härkönen L, Huttunen M et al (2020) Drainage for forestry increases N, P and TOC export to boreal surface waters. Sci Total Environ 762:144098. https://doi.org/10.1016/j.scitotenv.2020.144098

Finnish Forest Statistics (2019) Eds: Peltola A, Ihalainen A, Mäki-Simola E, Peltola A, Sauvula-Seppälä T, Torvelainen J, Uotila E, Vaahtera E, Ylitalo E. Natural Resources Institute Finland. Helsinki, Finland, 198 pp. https://stat.luke.fi/sites/default/files/suomen_metsatilastot_2019_verkko2.pdf. Accessed 27 May 2021

Finnish Forest Statistics (2020) Eds: Peltola A, Räty M, Sauvula-Seppälä T, Torvelainen J, Uotila E, Vaahtera E, Ylitalo E. Natural Resources Institute Finland. Helsinki, Finland, 198 pp. https://stat.luke.fi/sites/default/files/suomen_metsatilastot_2020_verkko.pdf. Accessed 27 May 2021

Gundersen P, Thybring EE, Nord-Larsen T, Vesterdal L, Nadelhoffer KJ, Johannsen VK (2021) Old-growth forest carbon sinks overestimated. Nature 591:E21–E23. https://doi.org/10.1038/s41586-021-03266-z

Heinonen T, Pukkala T, Ikonen V-P, Peltola H, Venäläinen A, Dupont S (2009) Integrating the risk of wind damage into forest planning. For Ecol Manag 258(7):1567–1577. https://doi.org/10.1016/j.foreco.2009.07.006

Heinonen T, Pukkala T, Mehtätalo L, Asikainen A, Kangas J, Peltola H (2017) Scenario analyses on the effects of harvesting intensity on development of forest resources, timber supply, carbon balance and biodiversity of Finnish forestry. Forest Policy Econ 80:80–98. https://doi.org/10.1016/j.forpol.2017.03.011

Heinonen T, Pukkala T, Kellomäki S, Strandman H, Asikainen A, Venäläinen A, Peltola H (2018) Effects of forest management and harvesting intensity on the timber supply from Finnish forests in a changing climate. Can J For 48:1–11. https://doi.org/10.1139/cjfr-2018-0118

Henttonen HM, Nöjd P, Mäkinen H (2017) Environment-induced growth changes in the Finnish forests during 1971–2010 – an analysis based on National Forest Inventory. For Ecol Manag 386:22–36. https://doi.org/10.1016/j.foreco.2016.11.044

Honkaniemi J, Lehtonen M, Väisänen H, Peltola H (2017) Effects of wood decay by *Heterobasidion annosum* on the vulnerability of Norway spruce stands to wind damage: a mechanistic modelling approach. Can J For 47(6):777–787. https://doi.org/10.1139/cjfr-2016-0505

Hurmekoski E, Myllyviita T, Seppälä J, Heinonen T, Kilpeläinen A, Pukkala T et al (2020) Impact of structural changes in wood-using industries on net carbon emissions in Finland. J Ind Ecol 24(4):899–912. https://doi.org/10.1111/jiec.12981

Hynynen J, Salminen H, Ahtikoski A, Huuskonen S, Ojansuu R, Siipilehto J, Eerikäinen K (2015) Long-term impacts of forest management on biomass supply and forest resource development: a scenario analysis for Finland. Eur J For Res 134(3):415–431. https://doi.org/10.1007/s10342-014-0860-0

Kalliokoski T, Bäck J, Boy M, Kulmala M, Kuusinen N, Mäkelä A et al (2020) Mitigation impact of different harvest scenarios of Finnish forests that account for albedo, aerosols, and trade-offs of carbon sequestration and avoided emissions. Front For Glob Change 3:562044. https://doi.org/10.3389/ffgc.2020.562044

Kärkkäinen L, Haakana M, Heikkinen J, Helin J, Hirvelä H, Jauhiainen L et al (2019) Maankäyttösektorin toimien mahdollisuudet ilmastotavoitteiden saavuttamiseksi Valtioneuvoston selvitys- ja tutkimustoiminnan julkaisusarja 67/2018 (in Finnish). http://urn.fi/URN:ISBN:978-952-287-618-8. Accessed 27 May 2021

Kauppi P, Hanewinkel M, Lundmark L, Nabuurs G-J, Peltola H, Trasobares A, Hetemäki L (2018) Climate smart forestry in Europe. European Forest Institute. https://efi.int/sites/default/files/files/publication-bank/2018/Climate_Smart_Forestry_in_Europe.pdf. Accessed 27 May 2021

Kellomäki S, Peltola H, Nuutinen T, Korhonen KT, Strandman H (2008) Sensitivity of managed boreal forests in Finland to climate change, with implications for adaptive management. Philos Trans R Soc B Biol Sci 363(1501):2341–2351. https://doi.org/10.1098/rstb.2007.2204

Kellomäki S, Strandman H, Heinonen T, Asikainen A, Venäläinen A, Peltola H (2018) Temporal and spatial change in diameter growth of boreal scots pine, Norway spruce and birch under recent-generation (CMIP5) global climate model 4 projections for the 21st century. Forests 9(3):118. https://doi.org/10.3390/f9030118

Kellomäki S, Väisänen H, Kirschbaum MUF, Kirsikka-Aho S, Peltola H (2021) Effects of different management options of Norway spruce on radiative forcing through changes in carbon stock and albedo. For Int J For Res cpab010. https://doi.org/10.1093/forestry/cpab010

Laapas M, Lehtonen I, Venäläinen A, Peltola H (2019) 10-year return levels of maximum wind speeds under frozen and unfrozen soil forest conditions in Finland. Climate 7(5):62. https://doi.org/10.3390/cli7050062

Lehtonen I, Kämäräinen M, Gregow H, Venäläinen A, Peltola H (2016a) Heavy snow loads in Finnish forests respond regionally asymmetrically to projected climate change. Nat Hazards Earth Syst Sci 16:2259–2271. https://doi.org/10.5194/nhess-16-2259-2016

Lehtonen I, Venäläinen A, Kämäräinen M, Peltola H, Gregow H (2016b) Risk of large-scale forest fires in boreal forests in Finland under changing climate. Nat Hazards Earth Syst Sci 16:239–253. https://doi.org/10.5194/nhess-16-239-2016

Lehtonen I, Venäläinen A, Kämäräinen M, Asikainen A, Laitila J, Anttila P, Peltola H (2019) Projected decrease in wintertime bearing capacity on different forest and soil types in Finland under a warming climate. Hydrol Earth Syst Sci 23:1611–1631. https://doi.org/10.5194/hess-23-1611-2019

Lehtonen A, Aro L, Haakana M, Haikarainen S, Heikkinen J, Huuskonen S et al (2021) Maankäyttösektorin ilmastotoimenpiteet: Arvio päästövähennysmahdollisuuksista. Luonnonvara- ja biotalouden tutkimus 7/2021 (In Finnish). Natural Resources Institute Finland. Helsinki, Finland, 121 pp. http://urn.fi/URN:ISBN:978-952-380-152-3. Accessed 27 May 2021

Leppä K, Hökkä H, Laiho R, Launiainen S, Lehtonen A, Mäkipää R et al (2020a) Selection cuttings as a tool to control water table level in boreal drained peatland forests. Front Earth Sci 8:576510. https://doi.org/10.3389/feart.2020.576510

Leppä K, Korkiakoski M, Nieminen M, Laiho R, Hotanen J-P, Kieloaho A-J et al (2020b) Vegetation controls of water and energy balance of a drained peatland forest: responses to alternative harvesting practices. Agric For Meteorol 295:108198. https://doi.org/10.1016/j.agrformet.2020.108198

Mäkelä H, Venäläinen A, Jylhä K, Lehtonen I, Gregow H (2014) Probabilistic projections of climatological forest fire danger in Finland. Clim Res 60:73–85. https://www.jstor.org/stable/24896175

Minkkinen K, Ojanen P, Koskinen M, Penttilä T (2020) Nitrous oxide emissions of undrained, forestry-drained, and rewetted boreal peatlands. For Ecol Manag 478:118494. https://doi.org/10.1016/j.foreco.2020.118494

Ojanen P, Penttilä T, Tolvanen A, Hotanen J-P, Saarimaa M, Nousiainen H, Minkkinen K (2019) Long-term effect of fertilization on the greenhouse gas exchange of low-productive peatland forests. For Ecol Manag 432:786–798. https://doi.org/10.1016/j.foreco.2018.10.015

Reyer C, Bathgate S, Blennow K, Borges JG, Bugmann H, Delzon S et al (2017) Are forest disturbances amplifying or cancelling out climate change-induced productivity changes in European forests? Environ Res Lett 12(3):034027. https://doi.org/10.1088/1748-9326/aa5ef1

Ruosteenoja K, Jylhä K, Kämäräinen M (2016) Climate projections for Finland under the RCP forcing scenarios. Geophysica 51: 17–50. http://www.geophysica.fi/pdf/geophysica_2016_51_1-2_017_ruosteenoja.pdf. Accessed 27 May 2021

Ruosteenoja K, Markkanen T, Venäläinen A, Räisänen P, Peltola H (2018) Seasonal soil moisture and drought occur - rence in Europe in CMIP5 projections for the 21st century. Clim Dyn 50:1177–1192. https://doi.org/10.1007/s00382-017-3671-4

Seppälä J, Heinonen T, Pukkala T, Kilpeläinen A, Mattila T, Myllyviita T et al (2019) Effect of increased wood harvesting and utilization on required greenhouse gas displacement factors of wood-based products and fuels. J Environ Manag 247:580–587. https://doi.org/10.1016/j.jenvman.2019.06.031

Statistics Finland (2020) Suomen kasvihuonekaasupäästöt 1990–2019. Helsinki, Finland (in Finnish), 83 pp. https://stat.fi/static/media/uploads/tup/khkinv/yymp_kahup_1990-2019_2020.pdf. Accessed 27 May 2021

Suomen ilmastopaneeli (2021) Ilmastolakiin kirjattavat pitkän aikavälin päästö- ja nielutavoit-
teet – Ilmastopaneelin analyysi ja suositukset. Suomen ilmastopaneelin raportti 1/2021 (in
Finnish), 13 p. https://www.ilmastopaneeli.fi/wp-content/uploads/2021/02/ilmastopaneelin-
raportti_ilmastolain-suositukset_final.pdf. Accessed 27 May 2021
Venäläinen A, Lehtonen I, Laapas M, Ruosteenoja K, Tikkanen O-P, Viiri H et al (2020) Climate
change induces multiple risks to boreal forests and forestry in Finland: a literature review. Glob
Change Biol 26:4178–4196. https://doi.org/10.1111/gcb.15183

Chapter 12
Climate-Smart Forestry Case Study: Germany

Marc Hanewinkel, Andrey Lessa Derci Augustynczik, and Rasoul Yousefpour

Abstract Forests cover approximately one-third of Germany's territory. They are among the most productive forests in Europe and in a position to contribute considerably to climate change mitigation. Germany has set national targets for climate mitigation via forests and measures such as conversion towards mixed and climate-adapted forests; a stronger control on the sustainability of imported solid biofuels; an increase in forest area; a reduction in the emissions related to forest soils, especially on drained peatlands; and a reduction in land take to less than 30 ha day^{-1}. Climate change is already exerting severe economic, environmental and social impacts on German forests and the forest-based sector, and this trend is likely to continue and intensify in the future. The key question for future is: how best to optimise the mitigation potential of the forests while at the same time adapting the forests to deal with ongoing climate change. This situation calls for a very careful balancing of strategies and a holistic approach, which the CSF framework can provide. Our simulation indicated that the opportunity costs of using high-valued and productive species, such as Norway spruce, for mitigation purposes (i.e. by the in-situ accumulation of carbon) produces high opportunity costs, while species of less value, such as European beech, would be better suited for this purpose. In order to follow a systematic approach combining mitigation and adaptation, we propose a generic framework for adaptation that takes into account the cost efficiency of all measures, and includes this in suggesting the most efficient ways to increase the mitigation potential of the forests in Germany. Current and emerging forest bio-economy products also offer significant potential for the future mitigation potential via substitution and carbon storage.

Keywords Climate-Smart Forestry · Mitigation-adaptation generic framework · Cost-efficiency

M. Hanewinkel (✉) · A. Lessa Derci Augustynczik · R. Yousefpour
University of Freiburg, Freiburg, Germany
e-mail: marc.hanewinkel@ife.uni-freiburg.de

© The Author(s) 2022
L. Hetemäki et al. (eds.), *Forest Bioeconomy and Climate Change*, Managing
Forest Ecosystems 42, https://doi.org/10.1007/978-3-030-99206-4_12

12.1 Climate Change and the Forest Sector in Germany

Climate change is exerting unprecedented pressure on German forests and their capacity to deliver ecosystem services, with an increase in drought, bark beetle and wind damage during the last few years. This poses a major challenge for the management of forest resources, where we need to not only balance the provisioning of different ecosystem services, but also anticipate and adapt to future climate impacts, and enhance the role of forest ecosystems in climate-mitigation portfolios. This is a difficult task, since the multiple uses of forests may give rise to trade-offs that need to be resolved, especially those concerning wood production goals, biodiversity conservation and climate protection (German Advisory Council on Global Change [WBGU] 2020).

Climate-smart forestry (CSF) has recently emerged as a framework for tackling these issues and enhancing the mitigation potential of forest ecosystems, while acknowledging the effects of climate change on forest dynamics and taking action to overcome these. Simultaneously, it considers the socioeconomic aspects of forest management (Nabuurs et al. 2017). In this context, policy and management must seek efficient solutions for the forest sector in order to mitigate climate change. The promotion of more-stable species compositions and higher structural diversity via modifications to the thinning and harvesting regimes have been proposed as a way of coordinating the adaptation and mitigation role of forest ecosystems in climate-mitigation efforts (Verkerk et al. 2020). Similarly, a focus on policies that promote the substitution effects of fossil-intensive materials by wood products has been viewed as a cost-effective way towards climate neutrality.

Forests cover approximately one-third of Germany's territory (Table 12.1). German forests are among the most productive forests in Europe, with an average annual increment of 11.2 m^3 ha^{-1} $year^{-1}$, and are thus in a position to contribute considerably to mitigation as part of a CSF approach. For example, in the period from 2002 to 2012, an average of around 76 million m^3/year were harvested in German forests. This is clearly below the average annual increment, highlighting the potential for an increase in the mitigation potential via carbon storage in wood products and substitution effects.

Germany has set national targets for climate mitigation via forest ecosystems. Among the milestones proposed for the next decade are: a conversion towards mixed and climate-adapted forests; a stronger control on the sustainability of imported solid biofuels; an increase in forest area; a reduction in the emissions related to forest soils, especially on drained peatlands; and a reduction in land take to less than 30 ha day^{-1}. The forest-based sector in Germany can contribute to climate mitigation via three channels—forest sinks, substitution and storage (in wood products). Given this, it is important to consider how these channels could be used to increasingly contribute to climate mitigation, as well as simultaneously adapting the forests to the changing climate.

Here, we address the issues introduced above, starting with an outline of the main impacts of climate change on German forests, including the effects of

Table 12.1 Overview of the forest sector in Germany

Forest resources	
Area	11.42 million ha
Growing stock	336 m³ ha⁻¹
Carbon storage	1.2 billion t C
Species composition	25% Norway spruce, 22% scots pine, 15% European beech, 10% oak, 20% other broadleaves, 3% other conifers
Ownership	
Private	50%
State	30%
Community	20%
Economic contribution	
Gross added value	€34.3 billion in 2017
Employment	1 million people
Companies	121,900 companies

Sources: Bundeswaldinventur (BWI 2012), Clusterstatistik Forst und Holz (Becher 2016)

changing precipitation regimes, temperatures and increased disturbance activity. Subsequently, we employ a simulation-optimisation model to assess the potential mitigation of German forests under climate change up to the end of the century, using a process-based model. We further discuss climate adaptation approaches and the options available to deal with the impacts of climate change. Finally, we outline future measures for the forest sector in Germany that will increase its climate-mitigation potential while observing the need to adapt forests to future climatic conditions.

12.2 Impacts of Climate Change

Climate change and its impacts on forests has been an important topic for practitioners and scientists for more than two decades. Due to the federal system in Germany, however, a consistent and unified strategy for the whole country has never been implemented. Instead, the different states (i.e. Bundesländer) have come up with individual approaches to deal with the ever-increasing, mostly negative impacts of climate change on forests, such as the biotic and abiotic disturbances that have, in recent years, reached critical levels across the whole country (see below).

Bolte et al. (2009) analysed the need and strategies for forest management measures to enhance adaptation to climate change in the German states. The important role of increasing biotic threats has been acknowledged by all stakeholders. Only slight differences have been found regarding tree species' adaptive potential to climate change, with Norway spruce expected to have a low adaptive potential, while introduced species, such as Douglas fir and red oak, have been assumed to be more adaptive. Several native species, such as European beech, have been considered as being quite tolerant in the face of climate change effects. The most obvious differences detected were regarding adaptation strategies. While some states have preferred active adaptation (e.g. forest transformation aimed at replacing sensitive tree species), others have come out more in favour of a combination of active adaptation and risk minimisation strategies (e.g. by establishing tree species mixtures). Passive adaptation has predominantly been a less-preferred option.

Since Bolte et al.'s (2009) assessment, the situation in Germany's forests has—similarly to the rest of Europe—dramatically changed. A series of storms, extreme drought events and bark-beetle attacks, in addition to forest fires in 2018–2020, have led to the greatest amount of forest damage in Germany since World War II. The Federal Ministry of Agriculture and Consumer Protection (BMEL) has estimated that, in 2018–2020, around 180 million m³ of wood have been damaged, and an area of 285,000 ha has had to be reforested (BMEL 2020/Internet source no. 3). Most of the damage is directly associated with the impacts of climate change. Norway spruce and Scots pine are the tree species that have been most affected, but native species, such as European beech and silver fir, have also exhibited severe problems. This casts doubt on the hypothesis that a more 'natural' forest composition, in terms of species, would significantly increase the resistance and resilience of forests in Germany under climate change. In 2020, the German government assigned the record sum of almost €550 million to support forest owners in dealing with the damage. Discussion on how to best use these funds is ongoing.

12.3 Economic Implications and the Potential for Climate-Change Mitigation

12.3.1 Economic Costs

Hanewinkel et al. (2010) calculated the economic effects of a predicted climate-change-induced shift from Norway spruce to European beech in a forest area of 1.3 million ha in southwestern Germany. The predicted shift led to a reduction in the potential area of Norway spruce by between 190,000 and 860,000 ha. The financial effect of this reduction on the land expectation value was estimated to be between €690 million and €3.1 billion. Using a similar methodological approach (see Hanewinkel et al. 2013), the total loss in German forest area (11 million ha) would equate to around €11 billion. This figure is, of course, subject to considerable uncertainty, yet it shows that the economic impacts of climate change on forests can be severe. In Germany, the timber industry relies, to a large degree, on coniferous species. These tree species are especially vulnerable to climate change, suggesting that the forest industry may be at significant risk in the future.

Bösch et al. (2017, 2019) assessed the costs and carbon sequestration potential of selected forest management measures in Germany, including the effects on the harvested-wood products pool, within a framework that accounted for both the financial impacts on the downstream industries and those on the values of non-market goods and forest services. They showed that these costs could amount to several billion euros per year, and that the cost-effectiveness could be very low. That is, the abatement costs per ton of CO_2 may be very high due to the high environmental costs.

12.3.2   Potential for Mitigation

Germany's forests are considered to be an important part in the climate-change mitigation strategy of the country. However, similarly to the forests in the EU (Nabuurs et al. 2017), their potential in that respect is still underused.

According to Wissenschaftlicher Beirat (2016), the annual potential of forests in Germany to mitigate greenhouse gases through sequestration and the substitution effects of wood products is estimated to be 127 Mt. CO_2eq. This is equal to 16% of Germany's greenhouse gas emissions in 2019 (805 Mt. CO_2eq.), a figure that is slightly higher than the one reported by Nabuurs et al. (2017) for the EU. Indeed, Wissenschaftlicher Beirat (2016) considered forests to be one of the most efficient terrestrial sinks in Germany.

Here, we employed a simulation-optimisation model, developed by Yousefpour et al. (2018), to assess the potential for increasing carbon sequestration in German forests along the lines of the CSF approach. We searched for optimal combinations of forest profitability (in terms of the net present value [NPV] of harvestings) and carbon sequestration *in situ*. We applied the approach to Germany under different climate-change scenarios, using a process-based forest-growth model to forecast future sequestration potential up to the end of the century. Subsequently, we identified management regimes that could realise these optimal combinations and assess the costs related to carbon sequestration. A central aspect of this analysis and CSF is the allocation of climate-mitigation actions to areas with simultaneously high sequestration potential and low opportunity costs, thereby increasing the efficiency of forestland use. In this sense, we also selected the species best suited for climate-mitigation actions. To this end, carbon sequestration was discounted in order to consider the urgency of the climate-mitigation actions using a 2% discount rate. Consequently, the carbon sequestration is expressed as present tons equivalent (PTE), meaning that 100 t of C sequestered 10 years in the future would represent 82 PTE. Wood-harvesting revenues were discounted using a 0.54% interest rate (Yousefpour et al. 2018).

As expected, forest profitability and carbon sequestration displayed a trade-off, since higher levels of carbon storage in the forests resulted in a decrease in wood utilisation and a reduction in harvesting revenues (Fig. 12.1). Therefore, forest owners applying climate-mitigation-oriented management would lose stumpage income (i.e. incur opportunity costs), which depends on the profitability of forest stands and their species composition. Figure 12.1a illustrates the total cost of generating

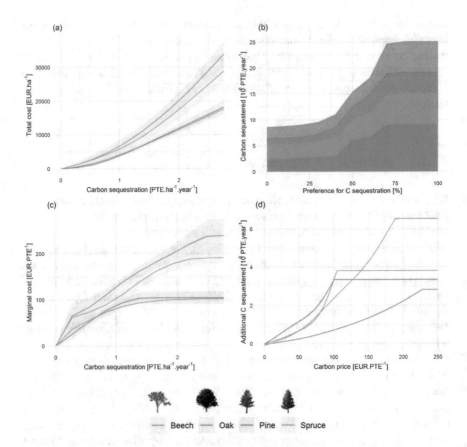

Fig. 12.1 Results of the multi-objective optimisation model for balancing forest profitability and carbon sequestration for the most abundant tree species in Germany. (**a**) Total cost incurred to increase carbon-sequestration (in PTE) levels *in situ* for the different species analysed. (**b**) Levels of carbon sequestration attained with different preferences for carbon sequestration over forest profitability. (**c**) Marginal cost curve for increasing carbon sequestration (i.e. the carbon supply curve). (**d**) Allocated sequestration potential for each species with increasing carbon price

additional carbon sequestration, in terms of NPV, compared to the baseline. For example, an additional sequestration of 2.5 PTE of C ha^{-1} $year^{-1}$ would require a compensation to forest owners in the range of €15–30 thousand ha^{-1}, with highest compensation required for oak and spruce forests. Beech and pine stands had the lowest compensation costs of about €15 thousand ha^{-1}. These patterns were maintained in the carbon supply curve for each species (Fig. 12.1c), which indicates the amount of increase in carbon sequestration that owners would be willing to adopt at different compensation levels (i.e. the carbon price), with the maximum compensation ranging from €100 to €239 PTE^{-1} for beech and oak stands, respectively.

Realisation of the maximum sequestration potential in Germany's forests may increase the carbon uptake nearly threefold compared to the baseline management

(Fig. 12.1b). Similarly to the supply function and total cost, different species displayed varying sequestration potential. This resulted from both the total areal share of the species and the biological potential (i.e. the carbon sequestration rates). Taking into account the maximum sequestration potential, spruce represented 36% of the total sequestration share, followed by pine (25%), beech (23%) and oak (16%) stands. Spruce stands had the largest share of forest area and growth rates, leading to higher sequestration levels. Despite the similar areal coverage, pine stands produced a smaller contribution, resulting from the lower growth rates of this species and the typically poorer sites it occupies, predominantly in sandy soils, which also have lower carbon storage capacity.

Considering that different species display diverging potentials to deliver ecosystem services, integrative approaches are required to ensure an efficient use of forest resources (WBGU 2020). For example, the promotion of mixed stands can balance the trade-offs related to production and climate-mitigation goals, as well as increase their resilience to disturbances and reduce the risks of catastrophic carbon losses (Jactel et al. 2017).

Figure 12.1d shows the share of the sequestration capacity realised with increasing compensation levels, up to the maximum potential. Hence, an earlier increase in carbon sequestration indicates a higher suitability of the species for increasing carbon sequestration. We found that pine and beech stands were preferable for mitigation actions, with a full allocation of the forests to carbon storage from compensations above €100 PTE^{-1}, and a slightly better suitability of pine stands. Conversely, spruce and oak stands were fully allocated to carbon sequestration actions only under high compensation payments of above €189 PTE^{-1} for the former and €228 PTE^{-1} for the latter.

The carbon sequestration levels computed here amounted to 4–11% of the country's carbon emissions in 2019–805 Mt. CO_2 (or approximately 8–23% if no carbon discounting is applied), which is compatible with previous estimates (Dunger et al. 2014). Therefore, forests may substantially contribute to the realisation of climate targets in the country.

The Wissenschaftlicher Beirat (2016) concluded that, in the agriculture and forestry sectors, forests play the most important role in carbon sequestration. In the package 'moderate climate protection', which predicts a mitigation effect of 65 Mt. CO_2eq./year, forests are expected to contribute 43% of the total (28 Mt. CO_2eq.). In the 'ambitious climate protection' case, forestry and HWPs contribute 56 Mt. CO_2eq. from a total of 130–135 Mt. CO_2eq. Adding this to the actual contribution of almost 130 Mt. from the forestry sector would make up for around 180 Mt. CO_2, thus amounting to an impact equal to 22% of the current level of yearly greenhouse gas emissions.

The potential to improve the mitigation effect of Germany's forests and the forest sector is of the same magnitude as Nabuurs et al. (2017) estimated for the impact of CSF measures on the whole of the EU (25%). However, it should be noted that the simulation analysis for Germany has several limitations. First, the potential to increase the forest area in Germany is naturally limited due to the high population density (250 people/km^2) and high pressure on land use for buildings and infrastructure, especially around the urban areas. Second, Germany already has a comparably high standing volume per hectare, limiting the increase in carbon in the living biomass. An Öko-Institut (2018) study indicated that the standing volume in Germany could be doubled, but we consider this to be unrealistic. Third, the potential to increase the standing volume by improved management through the conversion of coppice forests into high forests is limited, as coppices play virtually no role in management schemes in Germany. Fourth, Germany's forests are under increasing pressure from abiotic and biotic disturbances, as can be seen in the devastating drought and bark-beetle damage from 2018 to 2020. Hence, a high accumulation of biomass by CSF actions may increase the vulnerability of stands to windstorms and drought occurrences (e.g. Temperli et al. 2020). Similarly, fuel accumulation in unmanaged pine stands may pose the risk of wildfires, and extreme damage events in spruce stands may trigger the occurrence of bark-beetle outbreaks. The occurrence of such disturbance events could thus hinder climate-mitigation actions (Seidl et al. 2014).

12.4   The Role of Forest Products

The previous analysis only considered the role of forests in mitigation (and adaptation), but wood products can also play a significant role in the mitigation potential of the forest sector. Germany is a major producer of sawnwood and wood panels, and both the carbon storage in wood products and the substitution of wood for energy-intensive materials, especially in the construction sector, may contribute substantially to climate targets.

Bösch et al. (2017) estimated that wood substitution effects were up to the same order of magnitude—up to 18 Mt. CO_2 year^{-1}—as the forest carbon sink, depending on the wood utilisation scenario. It should be noted, however, that an increase in the mitigation potential associated with substitution effects was accompanied by a decrease in the sequestration potential of the forests due to the higher levels of wood removal.

Recent investments have been made in Germany to increase the production of new biomaterials and products that can replace fossil-based products. For example, UPM Biofuels has invested in a new biorefinery plant in the city of Leuna, with the capacity to produce 220,000 t of biochemicals annually (UPM Biofuels 2020). These biochemicals will enable a switch from fossil-based products to sustainable alternatives over a range of end uses, such as plastics, textiles, cosmetics and industrial applications. The plant is planned to start producing by the end of 2022. In 2020, the German government set out plans to accelerate its low-carbon transition by investing €3.6 billion in projects that help to strengthen its bioeconomy and create a market for bio-based products (https://biomarketinsights.com/germany-backs-e3-6bn-plan-to-support-bioeconomy-and-bio-based-products/).

An increase in the utilisation of wood products and improvements in the stewardship of wood imports are also predicted in the national climate action plan. The removal of barriers to the use of durable wood products (e.g. building regulations) and further investment in research and development towards the creation of new wood products are also being promoted, highlighting the importance of wood-based materials in the country's climate-mitigation portfolio. In addition, the WBGU (2020) has recommended boosting the use of timber in construction. According to the WBGU, timber from locally adapted, sustainable forestry offers effective possibilities for long-term carbon storage.

12.5   Nexus of Adaptation, Resilience and Mitigation: What Is the Right Way Forward?

Currently, because of the severe damage being done to forests in Germany, a public discussion has developed on how to manage forests under the impacts of climate change and how best to adapt the forests and increase their resistance and resilience to the changing environmental conditions. As this discussion has to do with the optimal strategy for combining mitigation and adaptation, it directly touches upon CSF.

It seems that two different groups have emerged, with fundamentally different approaches and opinions on how to manage forests under climate change. One approach, which you might call 'passive adaptation', is to keep the forests dense in order to maintain a cooler inner climate, and aims at a spontaneous adaptation using maximum natural processes, which opposes the classical forest management that has been practised over decades. This approach is supported by a highly diverse group, as well as certain specific regions (e.g. the Upper Rhine Valley). Some members of this group have expressed their opinions in an open letter to the Minister of Agriculture, thus putting pressure on politicians and bringing the case to a public debate. The alternative approach to this is what you might call the 'active adaptation approach', which aims at anticipating and adapting to the expected pressures posed by climate change. For example, promoting mixed forest stands, including the implementation of non-native species, and replanting large areas destroyed by drought and consecutive bark-beetle attacks with more resilient forest compositions. This active-adaptation approach has been supported by an official statement from the Scientific Board for forest policy of the Ministry of Agriculture and another official statement from the majority of German forest scientists (Deutscher Verband Forstlicher Forschungsanstalten [DVFFA] 2019).

The CSF (Chap. 9) approach combines all forest-based-sector mitigation possibilities (sink, substitution and storage) and adaption in a holistic way. Regarding adaptation, we point to a generic concept that has recently been developed (Yousefpour et al. 2017), which takes into account the cost efficiency (Fig. 12.2). According to this, a business-as-usual (BAU) strategy may still be the optimal choice if the cost of change (adaptation) exceeds the expected benefits. If the climate change impacts are low, a low-cost reactive adaptation may suffice. For scarce

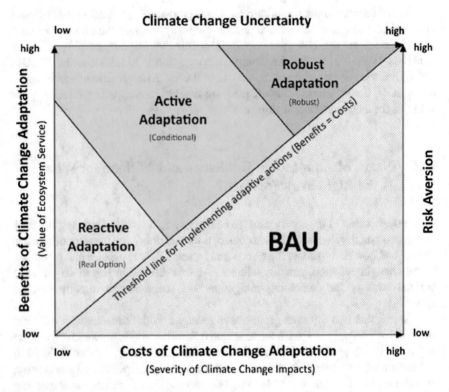

Fig. 12.2 Schematic allocation of different adaptation strategies in terms of costs and benefits under climate change. *BAU* business as usual

and valuable forest resources, especially under considerable climate change impacts, a proactive and robust strategy would be more suitable. The robust strategy is a more costly adaptation strategy, but it represents a better fit to the uncertainties inherent in climate change, and guarantees the provision of ecosystem services under all plausible climate-change scenarios.

Besides the importance of active adaptation, the Scientific Board of the Ministry of Agriculture (Wissenschaftlicher Beirat 2016) has proposed a list of measures for forestry and the forest sector to enhance their mitigation effect:

- Safeguard productive forests to sustainably use their potential for climate protection.
- Plant adapted and productive tree species, especially drought-tolerant conifers mixed with deciduous tree species.
- Increase the longevity of timber-based products and promote their cascade usage.
- Take into account climate protection effects when assigning protected areas in forests.
- Guarantee the protection of forest soils.
- Consult with and supervise small and medium private and community forest enterprises to reach climate protection goals.

- Communicate the positive climate protection services of forestry and the enhancement of timber usage.
- Giving up forestry and the harvesting of timber is not seen as an appropriate strategy for climate protection in the long run, although it may be an important instrument for achieving specific goals in biodiversity conservation.

From the perspective of the scale of the impact, the greatest mitigation potential in the forest-based sector would be achieved through:

- Changing the tree species composition in forestry production to generate more stable and resilient forests, capable of producing multiple benefits simultaneously, such as climate mitigation, wood production and habitat protection. This could even include an increase in coniferous species, but of course only mixed with native deciduous species (long-term effectivity).
- Protecting moors, inside and outside of forests (long-term effectivity).
- Producing lignocellulose from agricultural production, such as from short-rotation plantations (mid-term effectivity)
- Increasing the material usage of timber in long-lived timber products (long-term effectivity).

The Scientific Board estimated that the cost of avoiding greenhouse gas emissions in the forestry (and agricultural) sector will be dependent on the site and implementation of the measures, and that these are usually below €50/t CO_2eq.

12.6 Conclusions: Mitigation and Adaptation Go Hand in Hand

German forests are characterised by high standing volumes and productivity. They could play an important role in the overall potential of the country to mitigate climate change. Their potential to take up greenhouse gas emissions is in the range of, or even higher than, the average European forest. In addition, Germany is a major producer of sawnwood and wood-based panels, which also offers potential for climate mitigation, in terms of substituting for fossil-based materials and products and storing carbon in wood products.

On the other hand, climate change is already exerting severe economic, environmental and social impacts on German forests and the forest-based sector, and this trend is likely to continue and intensify in the future. There is a political debate taking place about how best to deal with this damage and minimise the risks in the future, asking, for example, how best to optimise the mitigation potential of the forests while at the same time adapting the forests to deal with ongoing climate change. This situation calls for a very careful balancing of strategies and a holistic approach, which the CSF framework can provide.

Our simulation indicated that the opportunity costs of using high-valued and productive species, such as Norway spruce, for mitigation purposes (i.e. by the *in-situ* accumulation of carbon) produces high opportunity costs, while species of less value, such as European beech, would be better suited for this purpose. In order to follow a systematic approach to addressing the challenges of combining mitigation and adaptation, we propose a generic framework for adaptation that takes into account the cost efficiency of all measures, and includes this in suggesting the most efficient ways to increase the mitigation potential of the forests in Germany. Current and emerging forest bioeconomy products also offer significant potential for the future mitigation potential via substitution and carbon storage.

References

Becher, G (2016) Clusterstatistik Forst und Holz: Tabellen für das Bundesgebiet und die Länder 2000 bis 2014. Braunschweig: Johann Heinrich von Thünen-Institut, 85 p, Thünen Working Paper 67, https://d-nb.info/1122831994

Bösch M, Elsässer P, Rock J, Rüter S, Weimar H, Dieter M (2017). Costs and carbon sequestration potential of alternative forest management measures in Germany. For Policy Econ 78:88–97. https://doi.org/10.1016/j.forpol.2017.01.005

Bösch M, Elsasser P, Rock J, Weimar H, Dieter M (2019). Extent and costs of forest-based climate change mitigation in Germany: accounting for substitution. Carbon Management 10(2):127–134

Bolte A, Eisenhauer D-R, Ehrhart H-P, Groß J, Hanewinkel M, Kölling C, Profft I, Rohde M, Röhe P, Amereller K (2009) Klimawandel und Forstwirtschaft – Übereinstimmungen und Unterschiede bei der Einschätzung der Anpassungsnotwendigkeiten und Anpassungsstrategien der Bundesländer. (Climate change and forest management – accordances and differences between the German states regarding assessments for needs and strategies towards forest adaptation). Landbauforschung VTI Agric For Res 4(59):269–278

Bundeswaldinventur (BWI) (2012) Die dritte Bundeswaldinventur. Inventur- und Auswertemethoden, 124. https://www.bundeswaldinventur.de/fileadmin/SITE_MASTER/content/Downloads/BWI_Methodenband_web.pdf

Deutscher Verband Forstlicher Forschungsanstalten (DVFFA) (2019) Neues Positionspapier: Anpassung der Wälder an den Klimawandel. Thünen Institut, Braunschweig

Dunger K, Stümer W, Oehmichen K, Riedel T, Ziche D, Grüneberg E, Wellbrock N (2014) Nationaler Inventarbericht Deutschland. Kapitel 7.2 Wälder. Umweltbundesamt (FEA). Clim Chang 24:524–571

Hanewinkel M, Hummel S, Cullmann D (2010) Modelling and economic evaluation of forest biome shifts under climate change in Southwest Germany. For Ecol Manag 259:710–719. https://doi.org/10.1016/j.foreco.2009.08.021

Hanewinkel M, Cullmann DA, Schelhaas MJ, Nabuurs G-J, Zimmermann NE (2013) Climate change may cause severe loss in the economic value of European forest land. Nat Clim Chang 3:203–207. https://doi.org/10.1038/nclimate1687

Jactel H, Bauhus J, Boberg J, Bonal D, Castagneyrol B, Gardiner B, Gonzalez JR, Koricheva J, Meurisse N, Brockerhoff EG (2017) Tree diversity drives forest stand resistance to natural disturbances. Curr For Rep 3(3):223–243

Nabuurs G-J, Delacote P, Ellison D, Hanewinkel M, Hetemäki L, Lindner M (2017) By 2050 the mitigation effects of EU forests could nearly double through climate smart forestry. Forests 8:484. https://doi.org/10.3390/f8120484

Öko-Institut (2018) Waldvision Deutschland. Beschreibung von Methoden, Annahmen und Ergebnissen – im Auftrag von Greenpeace. https://www.greenpeace.de/sites/www.greenpeace.de/files/publications/20180228-greenpeace-oekoinstitut-waldvision-methoden-ergebnisse.pdf. Accessed 10 Jan 2021

Seidl R, Schelhaas MJ, Rammer W, Verkerk PJ (2014) Increasing forest disturbances in Europe and their impact on carbon storage. Nat Clim Chang 4(9):806–810

Temperli C, Blattert C, Stadelmann G, Brändli UB, Thürig E (2020) Trade-offs between ecosystem service provision and the predisposition to disturbances: a NFI-based scenario analysis. For Ecosyst 7:1–17

UPM Biofuels (2020) The construction of UPM's innovative biochemicals facility starts in Germany. Press Release, 7 October 2020. https://www.upm.com/about-us/for-media/releases/2020/10/the-construction-of-upms-innovative-biochemicals-facility-starts-in-germany/. Accessed 10 Jan 2021

Verkerk PJ, Costanza R, Hetemäki L, Kubiszewski I, Leskinen P, Nabuurs GJ, Potocnik M, Palahí M (2020) Climate-smart forestry: the missing link. For Policy Econ 115:102164

WBGU (German Advisory Council on Global Change) (2020) Rethinking land in the Anthropocene: from separation to integration. Summary. WBGU, Berlin

Wissenschaftlicher Beirat Agrarpolitik, Ernährung und gesundheitlicher Verbraucherschutz und Wissenschaftlicher Beirat Waldpolitik beim BMEL (2016) Klimaschutz in der Land- und Forstwirtschaft sowie den nachgelagerten Bereichen Ernährung und Holzverwendung. Gutachten, Berlin

Yousefpour R, Augustynczik ALD, Hanewinkel M (2017) Pertinence of reactive, active and robust adaptation strategies in forest management under climate change. Ann For Sci 74:40. https://doi.org/10.1007/s13595-017-0640-3

Yousefpour R, Augustynczik ALD, Reyer C, Lasch P, Suckow F, Hanewinkel M (2018) Realizing mitigation efficiency of European commercial forests by climate smart forestry. Sci Rep 8:345. https://doi.org/10.1038/s41598-017-18778-w

Internet Sources

http://www.dvffa.de/system/files/files_site/Waldanpassung_Positionspapier%20des%20DVFFA_09_2019.pdf. Accessed 11 Nov 2020

https://bwi.info/. Accessed 12 Oct 2020

https://bwi.info/?inv=THG2017. Accessed 13 Oct 2020

https://www.bmel.de/DE/themen/wald/wald-in-deutschland/wald-trockenheit-klimawandel.html. Accessed 13 Oct 2020

https://www.bundesbuergerinitiative-waldschutz.de/unsere-positionen/offener-brief-an-bm-kl%C3%B6ckner/. Accessed 19 Oct 2020

https://www.upm.com/about-us/for-media/releases/2020/01/upm-invests-in-next-generation-biochemicals-to-drive-a-switch-from-fossil-raw-materials-to-sustainable-solutions/. Accessed 11 Sep 2020

VTI. https://www.thuenen.de/. Accessed 16 Oct 2020

Chapter 13
Climate-Smart Forestry Case Study: Spain

Elena Górriz-Mifsud, Aitor Ameztegui, Jose Ramón González, and Antoni Trasobares

Abstract In Spain, 55% of land area is covered by forests and other woodlands. Broadleaves occupy a predominant position (56%), followed by conifers (37%) and mixed stands (7%). Forest are distributed among the Atlantic (north-western Iberian rim), Mediterranean (rest of the peninsula including the Balearic Islands) and Macaronesian (Canary Islands) climate zones. Spanish woodlands provide a multiplicity of provisioning ecosystem services, such as, wood, cork, pine nuts, mushrooms and truffles. In terms of habitat services, biodiversity is highly relevant. Cultural services are mainly recreational and tourism, the latter being a crucial economic sector in Spain (including rural and ecotourism). Regulatory services, such as erosion control, water availability, flood and wildfire risk reduction, are of such great importance that related forest zoning and consequent legislation were established already in the eighteenth century. Climate change in Southern Europe is forecast to involve an increase in temperature, reduction in precipitation and increase in aridity. As a result, the risks for natural disturbances are expected to increase. Of these, forest fires usually have the greatest impact on ecosystems in Spain. In 2010–2019, the average annual forest surface area affected by fire was 95,065 ha. The combination of extreme climatic conditions (drought, wind) and the large proportion of unmanaged forests presents a big challenge for the future. Erosion is another relevant risk. In the case of fire, mitigation strategies should combine modification of the land use at the landscape level, in order to generate mosaics that will create barriers to the spread of large fires, along with stand-level prevention measures to either slow the spread of surface fires or, more importantly, impede the possibility of fire crowning or disrupt its spread. Similarly, forest management can play a major role in mitigating the impact of drought on a forest. According to the land use, land-use change and forestry (LULUCF) accounting, Spanish forests

E. Górriz-Mifsud (✉) · J. R. González · A. Trasobares
The Forest Science and Technology Centre of Catalonia (CTFC), Solsona, Spain
e-mail: elena.gorriz@ctfc.cat

A. Ameztegui
The Forest Science and Technology Centre of Catalonia (CTFC), Solsona, Spain

University of Lleida, Lleida, Spain

L. Hetemäki et al. (eds.), *Forest Bioeconomy and Climate Change*, Managing
Forest Ecosystems 42, https://doi.org/10.1007/978-3-030-99206-4_13

absorbed 11% of the total greenhouse gas emissions in 2019. Investments in *climate-smart forestry* provide opportunities for using all the different parts of the Spanish forest-based sector for climate mitigation—forest sinks, the substitution of wood raw materials and products for fossil materials, and the storage of carbon in wood products. Moreover, this approach simultaneously helps to advance the adaptation of the forest to changing climate and to build forest resilience.

Keywords Mediterranean forest · Wildfires · Forest bioeconomy · Non-wood forest products · Resilient landscapes · Unmanaged forests

13.1　Introduction to Spanish Forests and Their Utilisation

As in many other southern European regions, the land cover of Spain has changed considerably in the last century. The abandonment of a substantial proportion of rural activities in the primary sector since the 1960s has led to a progressive, spontaneous afforestation of many parcels. Consequently, Spain today has 55% of its land area covered by forests and other woodlands (Ministerio para la Transición Ecológica [MITECO] 2018). Broadleaves occupy a predominant position (56%), followed by conifers (37%) and mixed stands (7%) (Fig. 13.1). The most widespread forest formations are open forest (the *dehesas*), typically used for agroforestry, followed by Mediterranean oak (chiefly *Quercus ilex*) and pine

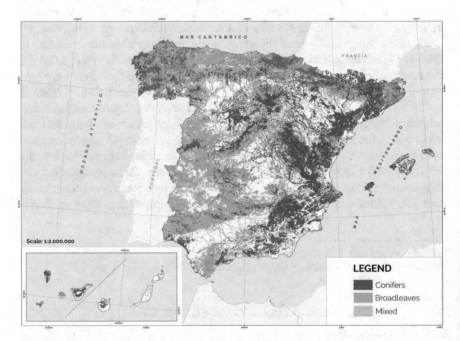

Fig. 13.1 Forest cover map of Spain. (Source: Ministerio de Medio Ambiente, Medio Rural y Marino, 2008)

(chiefly Aleppo pine). However, *Pinus pinaster* and *Pinus sylvestris* are the most important tree species in terms of timber volume. The rich ecosystem diversity is reflected in the more than 20 dominant tree species, which are distributed among the Atlantic (north-western Iberian rim), Mediterranean (rest of the peninsula including the Balearic Islands) and Macaronesian (Canary Islands) climate zones.

There is a large potential for increasing the use of domestic wood in Spain. Far from the typical European harvest rates, only one-third of the annual growing stock in Spain is harvested (Fig. 13.2) (Montero and Serrada 2013). While Spanish citizens tend to only consume small amounts of wood (0.8 m³/inhabitant/year – about half that of Central Europe and well below what Northern European countries consume), the aggregated annual timber consumption is almost double the domestic harvest. This means that, despite the available timber stock, over half of the demand needs to be covered by imported wood. The harvest intensity, however, varies considerably among autonomous communities, ranging from 10 to 30% in most Mediterranean regions, up to 60–70% in the Atlantic northern and north-western regions (with a maximum of 88% in Galicia).

Highly fragmented private parcels constitute most of the forest land (MITECO 2018), with more than 99% occupying less than 10 ha. Despite the clear management challenges this situation implies, over 80% of the wood harvest takes place on

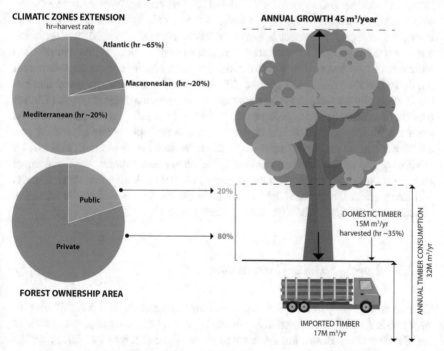

Fig. 13.2 Key Spanish forest and forestry data. (Source: Author elaboration based on MITECO 2018, Montero and Serrada 2013)

privately owned land, indicating that most productive forests (*Eucalyptus, Pinus radiata*, both of which are introduced species) tend to be owned by family forest owners. Only 18% of the forested land is subject to a management plan (MITECO 2018).

Beyond timber and fuelwood, Spanish forests also produce relevant non-timber forest products such as cork, pine nuts, chestnuts, resin, black truffles and wild mushrooms. Their value and markets are often imperfectly captured by the trade statistics, as are other ecosystem services (e.g. biodiversity conservation, water provision, amenities, carbon sequestration). Of the Spanish forests, 41% are nature-protection areas.

13.2 Impacts of Climate Change in Spanish Forests

13.2.1 Climate Change and Spanish Forests

Climate change in Southern Europe, and in Catalonia (north-eastern Spain) in particular, is forecast to involve an increase in temperature, reduction in precipitation and increase in aridity. Based on recent forest simulation studies (Trasobares et al. 2022; Morán-Ordóñez et al. 2020), the average annual mean temperature is projected to increase in Catalonia by 1.7–4.2 °C in this century, under Representative Concentration Pathways (RCPs) 4.5 and 8.5 (see Chap. 3). Consequently, climate change is expected to impact Spanish forests in several ways: (i) by decreasing water availability due to increased evapotranspiration due to the temperature increase; (ii) by increasing wildfire virulence as a result of reduced relative air humidity and increased wind speeds; (iii) by intensifying downpours, and increasing torrentiality and erosion-risk, especially in south-eastern Iberia and the Canary Islands, intimately linked to desertification; (iv) by increasing the frequency of wind storms, with stronger winds causing structural tree damage; (v) by expanding pest and disease areas and/or active periods due to reduced cold weather; and (vi) by modifying the phenology and physiology of plants and animals, with additional effects on biomass growth (Serrada Hierro et al. 2011). Altogether, these impacts will likely affect the current composition of forest species, as well as the provision of ecosystem services, while increasing forest risks.

13.2.2 Forest Species Composition

Climate change projections predict a significant contraction of the distribution of most mesic species in the Iberian Peninsula by 2100, but for widespread species in the Mediterranean Basin, the impact will be lessened (Lloret et al. 2013). In the mid-term (by 2040), in monospecific Catalan forests (Gil-Tena et al. 2019), a temperature increase of 1.2 °C (with a concomitant reduction in precipitation) may entail risk for *Pinus nigra*, *P. sylvestris*, *P. uncinata*, *Fagus sylvatica* and *Quercus pubescens*, while other tree species, such as *Pinus halepensis*, may have a lower risk. Forest stands in wetter and mountainous climatic sub-regions will attract higher risk than drier sub-regions, where *Pinus halepensis* prevails. This climatic risk will endanger the stand suitability of tree species that are less tolerant of drought conditions (i.e. causing a shift in tree species) and/or that have lower growth rates and a greater vulnerability to biotic hazards. Tree species dynamics are already showing rapid species shifts from conifers towards broadleaves (Vayreda et al. 2016), partly due to climatic variation, but also due to the legacy of human land use, mainly agricultural abandonment and reduced forest management intensity (e.g. coppicing for fuelwood). In some areas, tree species that used to be secondary are starting to become predominant. These changes also have economic consequences because the tree species that are becoming more common tend to have lower economic value in the markets.

13.2.3 Provisioning of Wood and Other Ecosystem Services

Spanish woodlands provide a multiplicity of ecosystem services (i.e. products), wood being the most relevant, followed by cork, pine nuts, mushrooms and truffles. In terms of habitat services, biodiversity is highly relevant. Cultural services are mainly recreational and tourism, the latter being a crucial economic sector in Spain (including rural and ecotourism). Regulatory services, such as erosion control, water availability, flood and wildfire risk reduction, are of such great importance that related forest zoning (*Montes de Utilidad Pública*) and consequent legislation were established as far back as the eighteenth century, and are still largely valid.

For some regions, in the short term, climate change may cause an increase in CO_2 sequestration and forest biomass productivity, such as in areas where water availability does not restrict growth, due to an increase in the vegetative period; this would benefit intensive silviculture (Serrada Hierro et al. 2011). On the other hand, a climate-sensitive forest scenario analysis conducted by Nabuurs et al. (2018), Morán-Ordóñez et al. (2020) and Trasobares et al. (2022) in north-eastern Spain indicated that, for the business-as-usual (BAU) scenario, climate change is expected to lead to denser forests with smaller tree diameter sizes, higher mortality rates and lower volume growth, and with a significantly greater risk of *forest fires* (see below). Morán-Ordóñez et al. (2020) found that the RCP8.5 scenario resulted in a decrease in all ecosystem services for all pine forests. The use of a BAU scenario with low-intensity harvesting resulted in the greatest *soil erosion mitigation* and *CO_2 storage*, but predicted lower (blue) water provision. Pardos et al. (2017) determined that, for Scots pine and Pyrenean oak forests in Valsaín (central Spain), wood production would decrease from 2060 onwards using the BAU. Nabuurs et al. (2018) and Trasobares et al. (2022) showed that the balance in net carbon emissions (also taking into account the life span of wood products, the substitution of fossil-based products, etc.) improved in management scenarios where *climate-smart forestry* and forest bioeconomy strategies were followed; that is, an increase in the managed area, improved silvicultural methods and incentivising the demand for construction timber in the medium term (2040–2050 onwards), with improved fire-risk prevention, drought and blue water provision in the shorter term.

The environmental conditions for cork oak have been predicted to decrease moderately in Andalucía under climate change. The risk will be more pronounced in the cork oaks planted in 1993–2000 as part of the EU's Rural Development Programme because many of these forests are located outside the optimal locations for these trees (Duque-Lazo et al. 2018a). Wild mushroom productivity may also be highly climate-dependent, with the extension of summer-like weather into the fruiting season (i.e. autumn) expected to diminish production. Surprisingly, Karavani et al. (2018a, b) predicted an increase in mushroom yield in pine forests in Catalonia under RCP4.5 and RCP8.5 during the twenty-first century. The autumn precipitation and soil moisture are expected to remain more or less stable (or even to increase slightly) during the fruiting season in 2016–2100, although temperatures are expected to increase compared to 2008–2015. This would mean the mushroom fruiting season would extend towards winter. Herrero et al. (2019) found consistent wild mushroom yields for *Pinus pinaster* in Castilla-y-León. Truffle productivity is also expected to shift under climate change, leading to lower-market-value species (summer truffles) becoming dominant relative to the current situation (Büntgen et al. 2012). Thomas and Büntgen (2019) also predicted a reduction in black truffle productivity in Spain due to climate change. Under RCP4.5, there could be an 88% harvest reduction due to increased summer temperatures and a 15.6% harvest reduction due to reduced summer precipitation. Under RCP8.5, there would be a total collapse in production. These effects could be at least partially overcome by the increased use of irrigation in specialised plantations.

Under climate change, *Pinus pinea* forests would have reduced pine nut yields (Pardos et al. 2015). Given that these forests are typically managed for pine cone productivity, future scenarios call for combining pine nuts with timber production. In terms of resin, the impact of climate change is still uncertain due to a lack of impact studies. However, based on our current understanding, a reduction in the tapping season is expected during the warmest months (June–September) (Rodríguez-García et al. 2015). Similarly, in years with a rainy summer and/or dry spring, a slightly longer tapping season might result, as resin yield increases after such events.

13.3 Forest Disturbances

13.3.1 Wildfires

The risks for abiotic (forest fires, erosion, drought, storms, etc.) and biotic (insects, disease) natural disturbances are expected to increase due to climate change (e.g. Seidl et al. 2014). Of these, forest fires usually have the greatest impact on ecosystems in Spain. In 2010–2019, the average annual forest surface area affected by fire was 95,065 ha (MITECO 2021a). The combination of extreme climatic conditions (drought, wind) and the large proportion of unmanaged forests presents a big challenge for the future. Erosion is another relevant risk. Most Spanish forests located on the steepest alpine and sub-alpine slopes are protected (Nabuurs et al. 2018).

Under climate change, extreme fire-weather conditions that can lead to large and catastrophic fires are expected to become more common (Piñol et al. 1998) as the number of extreme dry periods increases. Climate-change scenarios indicate an increase of 2–2.5 times the number of fires, 3.4–4.6 times the forest area burned, and 3–3.9 times the wooded area burned (Vázquez De La Cueva et al. 2012). An important aspect to consider is that the long-term impact on the vegetation or the adaptation of plants to fire does not depend on single events, but on fire regimes— that is, the fire characteristics for a given area over a certain period (Krebs et al. 2010). However, climate change is not the only factor that will modify the fire regimes on the Iberian Peninsula; other factors will define the size, frequency and/ or severity of the fires (Moreno et al. 2014). Moreover, changes in the fire activity have not been, and probably will not be, homogeneous over the Spanish territory. Past observations (Moreno et al. 2014) and future predictions (Jiménez-Ruano et al. 2020) have indicated that, in north-eastern Spain, there has been a general increase in fire activity both over an entire year and during the vegetative season, although this tendency is expected to decrease in the medium term (2036). On the other hand, in Spain overall, there is a trend towards fewer wildfires with lower intensities, and a reduction in the area burnt (MAPA 2019). This decrease can be attributed to improvements in, and expenditure on, fire suppression over the last few decades. However, even though past observations and future forecasts seem relatively optimistic, it is widely understood that the accumulation of fuel resulting from agricultural abandonment (Pausas and Paula 2012), areas of past fire exclusion (Piñol et al. 2005) and the expected increase in the number of days subject to extreme fire weather may lead to the unexpected occurrence of very large and catastrophic fires (Costa et al. 2011).

13.3.2 Water Scarcity and Drought

Interactions between the multiple drivers of global change can have diverse effects on the future condition of Mediterranean forests. Water scarcity will certainly be one of the most important agents of forest dynamics and their provision of services in the coming decades. The expected increase in evapotranspiration rates due to rising temperatures will come with a general reduction in water availability and greater precipitation irregularity, leading to more frequent, intense and prolonged droughts and hot spells. Many tree species in Spain will be particularly vulnerable to these events, including *Pinus sylvestris*, *Fagus sylvatica* and *Abies alba*. Decline in growth and increased die-back have already been reported in *Pinus sylvestris* populations in north-eastern Spain (Martínez-Vilalta and Piñol 2002) and in the southernmost populations of *Abies alba* in the Spanish Pyrenees (Macias et al. 2006). However, this phenomenon will not only affect the least-tolerant species–drought-adapted species are also likely to suffer the consequences of increased drought conditions. Drought has been linked to the general die-back of *Quercus ilex* in south-western Spain known as '*seca*', where weakened trees are more susceptible to attack by *Phytophthora* (Sánchez-Salguero et al. 2013). It has also been reported to cause growth decline in several pine species in south-eastern Spain (Sánchez-Salguero et al. 2012). We can expect a general reduction in site productivity in the medium and long terms, particularly in species or populations growing in water-limited environments, which includes most Iberian forests (Coll et al. 2021).

More importantly, we can expect different responses to disturbances across forest types. Evergreen gymnosperms growing in drought-prone areas have exhibited low resistance to, but faster recovery after, drought events compared to trees from temperate regions (Gazol et al. 2018). Therefore, the response of vegetation to changes in climate may be different as droughts become more intense and/or more frequent. This may ultimately affect forest compositions and species distributions. In the driest areas, desertification might advance and become a major problem (Karavani et al. 2018a).

Forest structure will also play a fundamental role in the response of the vegetation to drought, with dense, unmanaged forests being generally more vulnerable (Lindner and Calama 2013). Forests in dry areas are able to accommodate fewer trees per hectare for a given average size, and reduced stand density is known to increase drought resistance in several species (Martín-Benito et al. 2010). Earlier, more-intense thinnings have been proposed as a fundamental method in the toolkit of forest managers to help forests adapt to climate change (Vilà-Cabrera et al. 2018; Coll et al. 2021), constituting the basis of 'ecohydrological' or 'hydrology-oriented' silviculture (del Campo et al. 2017). Several modelling exercises have indeed suggested that intense reductions in stand density can help to reduce the impacts of climate change on stress-related mortality, particularly on xeric sites (Ameztegui et al. 2017).

13.3.3 Pests and Diseases

Climate change may affect the distribution of pathogens and hosts. Among the most relevant pests, the pine processionary moth causes most concern for conifer forests in Spain. It is expanding northwards and towards higher elevations due to milder winter conditions (Roques et al. 2015)—a trend shared across western Mediterranean Europe. This expansion may eventually accelerate the process of natural succession (i.e. the replacement of conifers by *Quercus* species), although higher rates of forest compositional change may be expected if more-destructive pest outbreaks than pine processionary moth occur (Gil-Tena et al. 2019). Imported pests, such as the pine nematode, entail additional relevant threats. Haran et al. (2015) indicated an expected expansion of the pine nematode towards higher altitudes, with the probability of it spreading into the Pyrenees, towards France and the rest of Europe.

In terms of disease, the pine pitch canker that affects *Pinus pinaster* and *Phytophthora cynnamomi* that mainly affects oaks can be highlighted. Serra-Varela et al. (2017) found that almost the entire Spanish distribution of *Pinus pinaster* will face an abiotic-driven exposure to pitch canker (due to the predicted increase in drought events under climate change), while the north-western edge of the Iberian Peninsula is predicted to face reduced exposure. Duque-Lazo et al. (2018b) indicated that oak decline provoked by *Phytophthora cynnamomi* may be reduced in Andalusian forests (southern Spain) until 2040, although the suitability of the habitat is predicted to increase after that.

13.4 Nexus for Adaptation and Resilience, and the Mitigation of Climate Change

13.4.1 Adaptation to Climate Change and Risk Management

Two of the most significant threats to Spanish forests, where the risk might be heightened in the future, are drought and fire. In the case of fire, it is widely recognised that mitigation strategies must be implemented at different scales (Gil-Tena et al. 2019). These should combine modification of the land use at the landscape level, in order to generate mosaics that will create barriers to the spread of large fires, along with stand-level prevention measures to either slow the spread of surface fires or, more importantly, impede the possibility of fire crowning or disrupt its spread (Loepfe et al. 2012). When implementing forest management interventions, it has been demonstrated that modifying the structure and composition of the forest at the stand level has an impact by reducing fire occurrence and damage (González et al. 2007). Consequently, specific management methods are being applied in certain regions of Spain (Piqué et al. 2017). It is clear that integrating these methods into the landscape, considering the spatial component of fire spread, has a much greater chance of mitigating the negative impacts of forest fires, or will facilitate the efficiency of suppression efforts, if specific measures are applied to high-priority areas (Gonzalez-Olabarria et al. 2019). Similarly, forest management can play a major role in mitigating the impact of drought on a forest (Martínez-Vilalta et al. 2012). Many of the management options considered to be appropriate for reducing competition for water resources (e.g. thinning) or for increasing the efficiency of the uptake and use of existing water (i.e. by favouring certain species admixtures based on their functional traits) (De Cáceres et al. 2021) may also be considered beneficial for reducing fire risk. The National Plan for Adaptation to Climate Change 2021–2031 actually considers these risks and mitigation goals as part of a broad, intersectoral plan (MITECO 2021b) and more-detailed forest-accountability plan (MITECO 2018).

13.4.2 The Role of Spanish Forests and Wood Products in Climate Change Mitigation

According to the land use, land-use change and forestry (LULUCF) accounting, Spanish forests absorbed 11% of the total greenhouse gas (GHG) emissions in 2019 – 314.529 Kt CO_2 eq. Table 13.1 details the impact of different forest subsector activities. The substitution of *forest biomass* for fossil-based energy in Spain is also important to take into account in this balance because of its potential and low cost (Turrado Fernández et al. 2016). The current energy consumption derived from biomass is close to 4 Mtoe. The 2030 bioenergy target of the National Integrated Plan for Energy and Climate indicates a need for an additional 1.6 Mtoe year^{-1} of electricity generation and 0.41 Mtoe year^{-1} for heating (MITECO 2020). These targets are perfectly achievable considering the estimated Spanish potential biomass for energy of 88.7 Mtoe year^{-1} (or 17.3 Mtoe year^{-1} for heating), with the portion coming from forests being 33.8 Mtoe year^{-1} (or 5.8 Mtoe year^{-1} for heating). This includes lumber industry residues, roundwood and other woody biomass from forestlands. Notably, the above figures do not take into account other potential sources, such as woody energy crops, and residues and side streams of the pulp and paper industry (Paredes-Sánchez et al. 2019). The use of timber in housing and construction (e.g. cross-laminated timber, plywood and sawn wood) is gaining more importance, although it is still far from reaching its potential use. Wood can store carbon for decades in buildings and can replace the use of fossil-intensive materials, such

Table 13.1 Contribution of Spanish forests and wood products to the GHG balance in 2019, and the forest reference levels (FRLs)

IPCC LULUCF sub-classes	GHG (Kt CO_2 eq.)	FRL 2021–2025
Forestland remaining as forestland	−29372.48	−29,303
Land converted to forestland	123.84	
Cropland converted to forestland	−2386.28	
Grassland converted to forestland	−1417.93	
Wetlands converted to forestland	−2.31	
Settlements converted to forestland	0.00	
Other land converted to forestland	−46.43	
Forestland converted to cropland	91.31	
Forestland converted to grassland	292.00	
Forestland converted to settlements	201.71	
Forestland converted to other land	0.00	
Harvested wood products	−2191.22	−1732
Wildfires (N$_2$O, CH$_4$)	Not available	330
Prescribed burning (N$_2$O, CH$_4$)	Not available	2
Forest contribution to the 2019 GHG balance	−34707.78	−30,703

IPCC Intergovernmental Panel on Climate Change
Source: MITECO (2018, 2021a)

as steel and concrete, therefore offering opportunities for climate mitigation in one of the most CO_2-intensive industry sectors.

Investments in *climate-smart forestry* provide opportunities for using all the different parts of the Spanish forest-based sector for climate mitigation—forest sinks, the substitution of wood raw materials and products for fossil materials, and the storage of carbon in wood products (Nabuurs et al. 2018). Moreover, this approach simultaneously helps to advance the adaptation of the forest to changing climate and to build forest resilience. The potential of *non-wood forest products* as substitutes for non-renewable materials has been poorly assessed so far. However, in terms of cork, Sierra-Pérez et al. (2018) comprehensively assessed its life-cycle for the purpose of building insulation. According to the study, using cork for insulation can have a positive CO_2 mitigation impact. The benefits are obvious when contrasted with mainstream, inorganic fibrous materials. Similarly, PricewaterhouseCoopers (PwC) and ECOBILAN (2008) found that the production of cork stoppers emitted less CO_2 (1.53 g CO_2/piece) than screw caps (37.17 g CO_2/piece) and synthetic caps (14.83 g CO_2/piece). When also accounting for the offsetting effect resulting from cork-oak forest management, cork stoppers become even more competitive (−113.2 g CO_2/piece).

13.4.3   Resilience of Spanish Forests

Many Mediterranean tree species have traits that give them the capacity to respond to the most frequent disturbances in an area—most notably, wildfires and drought events (response traits). However, because of the speed of current environmental change, the occurrence and severity of most disturbances has increased in forests across Europe (Senf and Seidl 2021). In the Mediterranean region, the severity, frequency and size of burned forest areas has increased over the last few decades (Turco et al. 2018), as have drought severity, heat waves and insect outbreaks (Balzan et al. 2020). The increasing frequency, size and severity of these disturbances will, in many cases, be beyond historical norms, and forests will likely often be overcome, particularly at the southern edges of their distributions (Vilà-Cabrera et al. 2012).

The concern about forest responses to disturbances has made resilience a new paradigm for researchers, managers and policy-makers. Considering resistance and resilience as two related, but distinct, components of ecosystem responses to disturbances, the resistance–resilience framework can provide a good understanding of post-disturbance forest dynamics (Sánchez-Pinillos et al. 2019), and may contribute to guiding climate-smart forestry and adaptive silviculture. Sánchez-Pinillos et al. (2016) developed the Persistence Index (PI) to assess the capacity of communities to maintain their functions and services following disturbances. The PI is based on the diversity, abundance and redundancy of response traits, under the assumption that an ecosystem will be more resilient and resistant to disturbances if it contains a greater share of species with a given set of traits that allow them to cope with disturbances. The application of the PI to Iberian forests highlights the importance of functional diversity rather than number of species as an indicator of forest resilience (Gazol et al. 2018). It can be used to operationalise the concepts of resistance and resilience in real-world management strategies, providing evidence for the adaptive management of forest ecosystems. However, vulnerability to disturbances can also vary along successional trajectories, which underscores the need to consider the temporal dimension in risk management.

Species-specific interactions may be altered under climate change and, according to the stress gradient hypothesis (Maestre et al. 2009), facilitative effects may become more frequent. The role of shrubs as nurse vegetation for pine seedlings has already been documented in semi-arid and arid Mediterranean regions (Gómez-Aparicio et al. 2008), but also in sub-Mediterranean pine woods (Sánchez-Pinillos et al. 2018). This role could become even more important in the future. The succession of disturbances may also impose a significant limitation on the resilience of forest stands. For example, the regeneration of *Pinus nigra* after wildfire depends both on the existence of nearby, unburned vegetation patches and on the climatic conditions in the years following the fire (Sánchez-Pinillos et al. 2018). The succession of fires and droughts, therefore, could trigger massive failures in regeneration, leading to a change in the ecosystem towards a greater dominance of oak. In the driest areas, the *combined effect of several disturbances* is likely to exceed the response capacity of the organisms, leading to the extinction of some species and even the disappearance of vegetation cover, which introduces a high risk for soil erosion, degradation and desertification.

13.5   Potential for a Forest-Based Bioeconomy in Spain

The Spanish forest sector accounted for 0.6% of the Gross Added Value in 2018, of which 0.9% came from forestry works, 0.19% from the timber and cork industry and 0.36% from the paper industry (INE 2021b). However, these figures do not consider the added value generated by most of wildfire management activities, hunting or forest foods (truffles, mushrooms, chestnuts, etc.), and therefore it clearly underestimates the total value of the forest-based sector. In 2011–2019, the Spanish forest sector employed about 130,000 people (INE 2021a).

Policies will be crucial for implementing a successful transition to a sustainable, circular bioeconomy and in contributing to the EU Green Deal Objectives in the coming decades. Policies such as the Next Generation Funds for COVID-19 recovery are supporting these objectives. For example, the funds include initiatives for increasing cross-laminated timber production, and the number of bioenergy plants and biorefineries. Spain's Bioeconomy Strategy 2015–2030 (Lainez et al. 2018) and the Climate Change Law 7/2021 provide incentives for moving to carbon neutrality, a necessary part of which will involve sustainable forest management and adapting forests to the changing climate.

Using forest biomass to replace fossil raw materials and products—the root cause of climate change—is essential. This implies increasing the use of forest biomass in, for example, the construction, packaging and textile sectors, and also for energy purposes, at least in the coming decade or two before other renewables (e.g. hydrogen) become more available. However, in Spain, forest management is the responsibility of the autonomous regions, and therefore it is crucial that they are ready to make the necessary changes at the regional level. Despite the large expansion of Spanish forestland in recent decades, the agricultural component of most bioeconomic initiatives is also important, and so it is necessary to advance and coordinate actions in both sectors. This is indeed being done, for example, in the Catalan Bioeconomy Strategy (2021–2030), the Basque Roadmap towards a Bioeconomy (2019), the Andalusian Circular Bioeconomy Strategy (2018), the Galician Agenda for the Forest Industry (2018), the recently established Research Centre for Rural Bioeconomy in Aragón, the CLAMBER project (Castilla–La Mancha Bio-Economy Region), and the Plan for Boosting Agro-food Bioeconomy in Castilla-y-León. The climate-smart forestry approach could play an important role in achieving the objectives of these strategies in the coming decades.

Acknowledgements We thank Jonas Oliva for his review of the biotic risks. Part of this chapter has been developed within the H2020 FIRE-RES project (Grant agreement ID: 101037419).

References

Ameztegui A, Cabon A, de Cáceres M, Coll L (2017) Managing stand density to enhance the adaptability of scots pine stands to climate change: a modelling approach. Ecol Model 356:141–150

Balzan M, Hassoun A, Aroua N, Baldy V, Bou Dagher M, Branquinho C, Dutay JC, El Bour M, Médail F, Mojtahid M, Morán-Ordóñez A, Roggero P, Rossi Heras S, Schatz B, Vogiatzakis I, Zaimes G, Ziveri P (2020) Ecosystems. In: Cramer W, Guiot J, Marini K (eds) Climate and environmental change in the Mediterranean Basin – current situation and risks for the future. First Mediterranean assessment report. Union for the Mediterranean, Plan Bleu, UNEP/MAP, p 151

Büntgen U, Egli S, Camarero JJ, Fischer EM, Stobbe U, Kauserud H, Tegel W, Sproll L, Stenseth NC (2012) Drought-induced decline in Mediterranean truffle harvest. Nat Clim Chang 2:827–829. https://doi.org/10.1038/nclimate1733

Coll L, Ameztegui A, Calama R, Lucas-Borja ME (2021) Dynamics and management of western Mediterranean pinewoods. In: Gidi N, Yagil O (eds) Pines and their mixed forest ecosystems in the Mediterranean Basin. Springer International Publishing, Switzerland

Costa P, Larrañaga A, Castellnou M, Miralles M, Kraus D (2011) Prevention of large wildfires using the fire types concept. European Forest Institute. ISBN 978-84-694-1457-6

De Cáceres M, Mencuccini M, Martin-StPaul N, Limousin JM, Coll L, Poyatos R, Cabon A, Granda V, Forner A, Valladares F, Martínez-Vilalta J (2021) Unravelling the effect of species mixing on water use and drought stress in Mediterranean forests: a modelling approach. Agric For Meteorol 296:108233. https://doi.org/10.1016/j.agrformet.2020.108233

del Campo AD, González-Sanchis M, Lidón A, García-Prats A, Lull C, Bautista I, Ruíz-Pérez G, Francés F (2017) Ecohydrological-based forest management in semi-arid climate. In: Křeček J, Haigh M, Hofer T, Kubin E, Promper C (eds) Ecosystem services of headwater catchments. Springer, pp 45–57. https://doi.org/10.1007/978-3-319-57946-7

Duque-Lazo J, Navarro-Cerrillo RM, Ruíz-Gómez FJ (2018a) Assessment of the future stability of cork oak (Quercus suber L.) afforestation under climate change scenarios in Southwest Spain. For Ecol Manag 409:444–456. https://doi.org/10.1016/j.foreco.2017.11.042

Duque-Lazo J, Navarro-Cerrillo RM, van Gils H, Groen TA (2018b) Forecasting oak decline caused by Phytophthora cinnamomi in Andalusia: identification of priority areas for intervention. For Ecol Manag 417:122–136. https://doi.org/10.1016/j.foreco.2018.02.045

Gazol A, Camarero JJ, Vicente-Serrano SM, Sánchez-Salguero R, Gutiérrez E, de Luis M, Sangüesa-Barreda G, Novak K, Rozas V, Tíscar PA, Linares JC, Martín-Hernández N, Martínez del Castillo E, Ribas M, García-González I, Silla F, Camisón A, Génova M, Olano JM, Longares LA, Hevia A, Tomás-Burguera M, Galván JD (2018) Forest resilience to drought varies across biomes. Glob Change Biol 24:2143–2158. https://doi.org/10.1111/gcb.14082

Gil-Tena A, Morán-Ordóñez A, Comas L, Retana J, Vayreda J, Brotons L (2019) A quantitative assessment of mid-term risks of global change on forests in Western Mediterranean Europe. Reg Environ Chang 19(3):819–831. https://doi.org/10.1007/s10113-018-1437-0

Gómez-Aparicio L, Zamora R, Castro J, Hódar JA (2008) Facilitation of tree saplings by nurse plants: microhabitat amelioration or protection against herbivores? J Veg Sci 19(2):161–172. https://doi.org/10.3170/2008-8-18347

González JR, Trasobares A, Palahí M, Pukkala T (2007) Predicting stand damage and tree survival in burned forests in Catalonia (North-East Spain). Ann For Sci 64(7):733–742. https://doi.org/10.1051/forest:2007053

Gonzalez-Olabarria JR, Reynolds KM, Larrañaga A, Garcia-Gonzalo J, Busquets E, Pique M (2019) Strategic and tactical planning to improve suppression efforts against large forest fires in the Catalonia region of Spain. For Ecol Manag 432:612–622. https://doi.org/10.1016/j.foreco.2018.09.039

Haran J, Roques A, Bernard A, Robinet C, Roux G (2015) Altitudinal barrier to the spread of an invasive species: could the Pyrenean chain slow the natural spread of the pinewood nematode? PLoS One 10(7):e0134126. https://doi.org/10.1371/journal.pone.0134126

Herrero C, Berraondo I, Bravo F, Pando V, Ordóñez C, Olaizola J, Martín-Pinto P, Oria de Rueda JA (2019) Predicting mushroom productivity from long-term field-data series in Mediterranean Pinus pinaster Ait. Forests in the context of climate change. Forests 10(3):1–18. https://doi.org/10.3390/f10030206

INE (Instituto Nacional de Estadística) (2021a) Empleados por rama actividad. Encuesta de Población Activa 2011–2020. https://www.ine.es/jaxiT3/Tabla.htm?t=4128. Accessed 23 Apr 2021

INE (2021b) Valor Agregado Bruto de las CNAE 02 16 17. Contabilidad Nacional de España 2011–2018. Agregados por ramas de Actividad. https://www.ine.es/jaxiT3/Tabla.htm?t=32449. Accessed 10 Apr 2021

Jiménez-Ruano A, de la Riva Fernández J, Rodrigues M (2020) Fire regime dynamics in mainland Spain. Part 2: a near-future prospective of fire activity. Sci Total Environ 705. https://doi.org/10.1016/j.scitotenv.2019.135842

Karavani A, Boer MM, Baudena M, Colinas C, Díaz-Sierra R, Pemán J, de Luis M, Enríquez-de-Salamanca Á, Resco de Dios V (2018a) Fire-induced deforestation in drought-prone Mediterranean forests: drivers and unknowns from leaves to communities. Ecol Monogr 88(2):141–169

Karavani A, De Cáceres M, Martínez de Aragón J, Bonet JA (2018b) Effect of climatic and soil moisture conditions on mushroom productivity and related ecosystem services in Mediterranean pine stands facing climate change. Agric For Meteorol 248:432–440. https://doi.org/10.1016/j.agrformet.2017.10.024

Krebs P, Pezzatti GB, Mazzoleni S, Talbot LM, Conedera M (2010) Fire regime: history and definition of a key concept in disturbance ecology. Theory Biosci 129(1):53–69. https://doi.org/10.1007/s12064-010-0082-z

Lainez M, González JM, Aguilar A, Vela C (2018) Spanish strategy on bioeconomy: towards a knowledge based sustainable innovation. New Biotechnol 40:87–95. https://doi.org/10.1016/j.nbt.2017.05.006

Lindner M, Calama R (2013) Mediterranean forests need to adapt to climate change. In: Lucas-Borja ME (ed), Forest Management of Mediterranean Forests under the new context of climate change: building alternatives for the coming future. NOVA science Publisher, pp 13–28

Lloret F, Martinez-Vilalta J, Serra-Diaz JM, Ninyerola M (2013) Relationship between projected changes in future climatic suitability and demographic and functional traits of forest tree species in Spain. Clim Chang 120(1–2):449–462. https://doi.org/10.1007/s10584-013-0820-6

Loepfe L, Martinez-Vilalta J, Piñol J (2012) Management alternatives to offset climate change effects on Mediterranean fire regimes in NE Spain. Clim Chang 115(3–4):693–707. https://doi.org/10.1007/s10584-012-0488-3

Macias M, Andreu L, Bosch O, Camarero JJ, Gutiérrez E (2006) Increasing aridity is enhancing silver fir (Abies alba Mill.) water stress in its south-western distribution limit. Clim Chang 79:289–313. https://doi.org/10.1007/s10584-006-9071-0

Maestre FT, Callaway RM, Valladares F, Lortie CJ (2009) Refining the stress-gradient hypothesis for competition and facilitation in plant communities. J Ecol 97(2):199–205. https://doi.org/10.1111/j.1365-2745.2008.01476.x

MAPA (Ministerio de Agricultura, Pesca y Alimentación) (2019) Los Incendios Forestales en España. Decenio 2006–2015. https://www.mapa.gob.es/es/desarrollo-rural/estadisticas/incendios-decenio-2006-2015_tcm30-511095.pdf. Accessed 20 Jan 2021

Martín-Benito D, del Río M, Heinrich I, Helle G, Cañellas I (2010) Response of climate-growth relationships and water use efficiency to thinning in a Pinus nigra afforestation. For Ecol Manag 259:967–975. https://doi.org/10.1016/j.foreco.2009.12.001

Martínez-Vilalta J, Piñol J (2002) Drought-induced mortality and hydraulic architecture in pine populations of the NE Iberian Peninsula. For Ecol Manag 161:247–256

Martínez-Vilalta J, Lloret F, Breshears DD (2012) Drought-induced forest decline: causes, scope and implications. Biol Lett 8(5):689–691. https://doi.org/10.1098/rsbl.2011.1059

MITECO (2018) Plan de Contabilidad Forestal Nacional. Plan de contabilidad forestal nacional para España, incluyendo el nivel forestal de referencia 2021–2025. https://www.miteco.gob.es/es/cambio-climatico/publicaciones/informes/nfap_es_tcm30-485874.pdf. Accessed 12 Apr 2021

MITECO (2020) Plan Nacional Integrado de Energía y Clima 2021–2030. https://www.miteco.gob.es/es/prensa/pniec.aspx. Accessed 12 Apr 2021

MITECO (Ministerio para la Transición Ecológica y el Reto Demográfico) (2021a) Incendios forestales. Avance informativo del 1 de enero al 31 de diciembre de 2020. https://www.mapa.gob.es/es/desarrollo-rural/estadisticas/avanceinformativoa31dediciembre2020_tcm30-132566.pdf. Accessed 13 Apr 2021

MITECO (2021b) Plan Nacional de Adaptación al Cambio Climático 2021–2030. https://www.miteco.gob.es/es/cambio-climatico/temas/impactos-vulnerabilidad-y-adaptacion/plan-nacional-adaptacion-cambio-climatico/default.aspx. Accessed 14 Apr 2021

MITECO (Ministerio para la Transición Ecológica) (2018) Avance Anuario Estadística Forestal 2018. https://www.mapa.gob.es/es/desarrollo-rural/estadisticas/aef_2018_resumen_tcm30-543072.pdf. Accessed 15 Apr 2021

Montero G, Serrada R (eds) (2013) La situación de los bosques y el sector forestal en España – ISFE 2013. Sociedad Española de Ciencias Forestales

Morán-Ordóñez A, Ameztegui A, De Cáceres M (2020) Future trade-offs and synergies among ecosystem services in Mediterranean forests under global change scenarios. Ecosyst Serv 45:101174. https://doi.org/10.1016/j.ecoser.2020.101174

Moreno MV, Conedera M, Chuvieco E, Pezzatti GB (2014) Fire regime changes and major driving forces in Spain from 1968 to 2010. Environ Sci Pol 37:11–22. https://doi.org/10.1016/j.envsci.2013.08.005

Nabuurs G, Verkerk PJ, Schelhaas MJ, González Olabarria JR, Trasobares A, Cienciala E (2018) Climate-smart forestry: mitigation impacts in three European regions. European Forest Institute

Pardos M, Calama R, Maroschek M, Rammer W, Lexer MJ (2015) A model-based analysis of climate change vulnerability of *Pinus pinea* stands under multiobjective management in the Northern Plateau of Spain. Ann For Sci 72:1009–1021. https://doi.org/10.1007/s13595-015-0520-7

Pardos M, Pérez S, Calama R, Alonso R, Lexer MJ (2017) Ecosystem service provision, management systems and climate change in Valsaín forest, Central Spain. Reg Environ Chang 17(1):17–32. https://doi.org/10.1007/s10113-016-0985-4

Paredes-Sánchez JP, López-Ochoa LM, López-González LM, Las-Heras-Casas J, Xiberta-Bernat J (2019) Evolution and perspectives of the bioenergy applications in Spain. J Clean Prod 213:553–568. https://doi.org/10.1016/j.jclepro.2018.12.112

Pausas JG, Paula S (2012) Fuel shapes the fire-climate relationship: evidence from Mediterranean ecosystems. Glob Ecol Biogeogr 21(11):1074–1082. https://doi.org/10.1111/j.1466-8238.2012.00769.x

Piñol J, Terradas J, Lloret F (1998) Climate warming, wildfire hazard, and wildfire occurrence in coastal eastern Spain. Clim Chang 38(3):345–357. https://doi.org/10.1023/A:1005316632105

Piñol J, Beven K, Viegas DX (2005) Modelling the effect of fire-exclusion and prescribed fire on wildfire size in Mediterranean ecosystems. Ecol Model 183:397–409. https://doi.org/10.1016/j.ecolmodel.2004.09.001

Piqué M, Vericat P, Beltrán M (2017) ORGEST: regional guidelines and silvicultural models for sustainable forest management. For Syst 26(2):1–6. https://doi.org/10.5424/fs/2017262-10627

PwC/ECOBILAN (2008) Evaluation of the environmental impacts of Cork Stoppers versus Aluminium and Plastic Closures. Analysis of the life cycle of Cork, Aluminium and Plastic Wine Closures. https://www.amorim.com/xms/files/v1/Sustentabilidade/Casos_de_Estudo/2008_-_LCA_Final.pdf. Accessed 11 May 2021

Rodríguez-García A, Antonio J, López R, Mutke S, Pinillos F, Gil L (2015) Influence of climate variables on resin yield and secretory structures in tapped *Pinus pinaster* Ait. in Central Spain. Agric For Meteorol 202:83–93

Roques A, Rousselet J, Avcı M, Avtzis DN, Basso A, Battisti A, Ben Jamaa ML, Bensidi A, Berardi L, Berretima W, Branco M, Chakali G, Çota E, Dautbasic M, Delb H, El Alao El Fels MA, El Mercht S, El Mokhefi M, Forster B, Garcia J, Georgiev G, Glavendekic MM, Goussard F, Halbig P, Henke L, Hernández R, Hódar JA, Ipekdal K, Jurc M, Klimetzek D, Laparie M, Larsson S, Mateus E, Matosevic D, Meier F, Mendel Z, Meurisse N, Mihajlovic L, Mirchev P, Nasceski S, Nussbaumer C, Paiva MR, Papazova I, Pino J, Podlesnik J, Poirot J, Protasov A, Rahim N, Sánchez Peña G, Santos H, Sauvard D, Schopf A, Simonato M, Tsankov G, Wagenhoff E, Yart A, Zamora R, Zamoum M, Robinet C (2015) Climate warming and past and present distribution of the processionary moths (*Thaumetopoea* spp.) in Europe, Asia Minor and North Africa. In: Roques A (ed) Processionary moths and climate change: an update. Ed. Quae, pp 81–162. ISBN 978-94-017-9339-1

Sánchez-Pinillos M, Coll L, De Cáceres M, Ameztegui A (2016) Assessing the persistence capacity of communities facing natural disturbances on the basis of species response traits. Ecol Indic 66:76–85. https://doi.org/10.1016/j.ecolind.2016.01.024

Sánchez-Pinillos M, Ameztegui A, Kitzberger T, Coll L (2018) Relative size to resprouters determines post-fire recruitment of non-serotinous pines. For Ecol Manag 429:300–307. https://doi.org/10.1016/j.foreco.2018.07.009

Sánchez-Pinillos M, De Cáceres M, Ameztegui A, Coll L (2019) Temporal dimension of forest vulnerability to fire along successional trajectories. J Environ Manag 248. https://doi.org/10.1016/j.jenvman.2019.109301

Sánchez-Salguero R, Navarro-Cerrillo RM, Camarero JJ, Fernández-Cancio Á (2012) Selective drought-induced decline of pine species in southeastern Spain. Clim Chang 113:767–785. https://doi.org/10.1007/s10584-011-0372-6

Sánchez-Salguero R, Camarero JJ, Dobbertin M, Fernández-Cancio Á, Vilà-Cabrera A, Manzanedo RD, Zavala MA, Navarro-Cerrillo RM (2013) Forest ecology and management contrasting vulnerability and resilience to drought-induced decline of densely planted vs . natural rear-edge *Pinus nigra* forests. For Ecol Manag 310:956–967. https://doi.org/10.1016/j.foreco.2013.09.050

Seidl R, Schelhaas M, Rammer W, Verkerk PJ (2014) Increasing forest disturbances in Europe and their impact on carbon storage. Nat Clim Chang 4: 806–810. https://doi.org/10.1038/nclimate2318.Increasing

Serrada Hierro R, Aroca Fernández MJ, Roig Gómez S, Bravo Fernández A, Gómez Sanz V (2011) Impactos, vulnerabilidad y adaptación al cambio climático en el sector forestal. Ministerio de Medio Ambiente y Medio Rural y Marino. https://www.miteco.gob.es/es/cambio-climatico/temas/impactos-vulnerabilidad-y-adaptacion/SECTOR%20FORESTAL_DOCUMENTO%20COMPLETO_tcm30-178472.pdf. Accessed 19 Apr 2021

Senf C, Seidl R (2021) Mapping the forest disturbance regimes of Europe. Nat Sustain 4: 63–70. https://doi.org/10.1038/s41893-020-00609-y

Serra-Varela MJ, Alía R, Pórtoles J, Gonzalo J, Soliño M, Grivet D, Raposo R (2017) Incorporating exposure to pitch canker disease to support management decisions of *Pinus pinaster* Ait. in the face of climate change. PLOS One:e0171549. https://doi.org/10.1371/journal.pone.0171549

Sierra-Pérez J, García-Pérez S, Blanc S, Boschmonart-Rives J, Gabarrell X (2018) The use of forest-based materials for the efficient energy of cities: environmental and economic implications of cork as insulation material. Sustain Cities Soc 37:628–636. https://doi.org/10.1016/j.scs.2017.12.008

Thomas P, Büntgen U (2019) A risk assessment of Europe's black truffle sector under predicted climate change. Sci Total Environ 655:27–34. https://doi.org/10.1016/j.scitotenv.2018.11.252

Trasobares A, Mola-Yudego B, Aquilué N, González-Olabarria JR, Garcia-Gonzalo J, García-Valdés R, De Cáceres M (2022) Nationwide climate-sensitive models for stand dynamics and forest scenario simulation. For Ecol Manag 505:119909. https://doi.org/10.1016/j.foreco.2021.119909

Turco M, Rosa-Cánovas JJ, Bedia J, Jerez S, Montávez JP, Llasat MC, Provenzale A (2018) Exacerbated fires in Mediterranean Europe due to anthropogenic warming projected with non-stationary climate-fire models. Nat Commun 9(1):1–9. https://doi.org/10.1038/s41467-018-06358-z

Turrado Fernández S, Paredes Sánchez JP, Gutiérrez Trashorras AJ (2016) Analysis of forest residual biomass potential for bioenergy production in Spain. Clean Techn Environ Policy 18(1):209–218. https://doi.org/10.1007/s10098-015-1008-8

Vayreda J, Martinez-Vilalta J, Gracia M, Canadell JG, Retana J (2016) Anthropogenic-driven rapid shifts in tree distribution lead to increased dominance of broadleaf species. Glob Change Biol 22(12):3984–3995. https://doi.org/10.1111/gcb.13394

Vázquez De La Cueva A, Quintana JR, Cañellas I (2012) Fire activity projections in the SRES A2 and B2 climatic scenarios in peninsular Spain. Int J Wildland Fire 21(6):653–665. https://doi.org/10.1071/WF11013

Vilà-Cabrera A, Rodrigo A, Martínez-Vilalta J, Retana J (2012) Lack of regeneration and climatic vulnerability to fire of scots pine may induce vegetation shifts at the southern edge of its distribution. J Biogeogr 39(3):488–496. https://doi.org/10.1111/j.1365-2699.2011.02615.x

Vilà-Cabrera A, Coll L, Martínez-Vilalta J, Retana J (2018) Forest management for adaptation to climate change in the Mediterranean basin: a synthesis of evidence. For Ecol Manag 407:16–22. https://doi.org/10.1016/j.foreco.2017.10.021

Chapter 14
The Way Forward: Management and Policy Actions

Lauri Hetemäki, Jyrki Kangas, Antti Asikainen, Janne Jänis, Jyri Seppälä, Ari Venäläinen, and Heli Peltola

> *For every complex problem there is an answer that is clear, simple, and wrong.*
>
> Henry L. Mencken.

Abstract Along with the evidence and analyses expounded on in this book, this chapter provides conclusions and suggestions concerning policy implications. These are based on a perspective that calls attention to the need for a holistic approach to look at the nexus of forests, the bioeconomy and climate change. Moreover, it is emphasised that, given the different uses of forests and the scarcity of forest resources, it makes sense to try to find ways to maximise synergies and minimise trade-offs between the different usages of forests. The forest-based sector contributes to climate-change mitigation via three channels—forests are a carbon sink, forest-based products can substitute for fossil-based products, and these products can store carbon for up to centuries. However, achieving these mitigation potentials in the future depends on forests being made resilient to the changing climate. Therefore, mitigation and adapting forests to climate change are married, both needing to be advanced simultaneously. Globally and in the EU, around 80–90% of the CO_2 emissions originate from the use of coal,

L. Hetemäki (✉)
European Forest Institute, Joensuu, Finland

Faculty of Agriculture and Forestry, University of Helsinki, Helsinki, Finland
e-mail: lauri.a.hetemaki@helsinki.fi

J. Kangas · J. Jänis · H. Peltola
University of Eastern Finland, Joensuu, Finland

A. Asikainen
Natural Resources Institute, Helsinki, Finland

J. Seppälä
Finnish Environment Institute, Helsinki, Finland

A. Venäläinen
Finnish Meteorological Institute, Helsinki, Finland

© The Author(s) 2022
L. Hetemäki et al. (eds.), *Forest Bioeconomy and Climate Change*, Managing Forest Ecosystems 42, https://doi.org/10.1007/978-3-030-99206-4_14

oil and natural gas. Consequently, the core issue in the fight against climate change is the phasing out of fossil-based products. Reaching this goal will not be possible without substituting also forest-based bioproducts for the purposes we are using oil, coal and gas for today. In the EU, this implies paying more attention to the need to develop new innovations in the forest bioeconomy, improve the resource efficiency and circularity of the bioproducts already available, and monitor the environmental sustainability of the bioeconomy.

Keywords Forest bioeconomy · Climate change · Adaptation to climate change · Holistic approach · Climate smart forestry · Science policy

14.1 The Nexus of Forests, the Bioeconomy and Climate Change

The quote preceding this chapter is fitting for the topic of this book—the nexus of forests, the bioeconomy and climate change. How are forests, the bioeconomy and climate change interlinked, and how do they impact on each other? As this book has demonstrated, the answers to these questions are characterised by complexity and a fair number of features that point even to wicked problems. When you first think you have found a clear and simple answer, a second thought reveals it to be only partially useful, or applicable only under a set of restrictive conditions or, in the worst case, simply wrong.

In this chapter, we provide insights and recommendations for policy actions. Along with the evidence and analyses expounded on in the previous chapters, these are also based on a perspective that emphasises the need for a *holistic approach* for viewing the nexus of forests, the bioeconomy and climate change. By this, we mean the following.

First, the approach is based on a self-evident, but often forgotten, fact. That is, forest resources are not limitless, but always scarce, despite being renewable. This is true even for the most forested country in the EU—Finland—where forests account for 74% of the land area. Moreover, there are multiple needs for forests and their use, such as providing raw materials, biodiversity, food (e.g. berries and mushrooms), recreation, hunting and carbon sequestration. Their importance has also evolved over time, especially in response to changing societal values, human needs, environmental change and technological development. For example, forest carbon sequestration has become a large societal need only in the last decade. The scarcity of forest resources relative to human need has always created potential trade-offs between the different uses of forests.

These facts bring to the fore the second most important feature of this book's approach. That is, given the different uses of forests and the scarcity of forest resources, it makes sense to try to find ways to maximise synergies and minimise trade-offs between the usages. Oftentimes these possibilities are not fully appreciated by people, policy-makers or even scientists, who may, for example, find the trade-offs between wood production and climate mitigation or between wood production and biodiversity inevitable. Therefore, one seems to have to choose an either/or. However, these trade-offs are not always inevitable and, in cases where these exist, there is usually the possibility of trying to minimise the trade-offs and maximise the synergies. This could be done e.g. using multi-objective forest management in which the simultaneous maximisation of multiple objectives increases the overall production levels of several ecosystem services (Biber et al. 2020; Díaz-Yáñez et al. 2020; Krumm et al. 2020). Indeed, it has even been argued that, in a modern society for example, biodiversity is necessary to the bioeconomy, and vice versa (Hetemäki et al. 2017; Palahi et al. 2020a, b). On the other hand, to achieve climate-change mitigation goals in the long term, forests should also be used for products that can substitute for fossil-based raw materials, the use of which is the root cause of climate change.

If one accepts the principle of these arguments, then the need to find synergies and minimise trade-offs between the bioeconomy, climate-change mitigation and biodiversity becomes a necessity. The downside of understanding this is that the world becomes much more complex. As a result, there is no longer any one single and simple solution to how the forest-based sector could, in the best possible way, contribute to climate-change mitigation or ensure biodiversity. Instead, diverse and tailored solutions are needed to accommodate different regions and circumstances. In this book, we have argued that *climate-smart forestry*, tailored to regional circumstances, provides a useful approach for increasing the forest-based sector's mitigation potential and helping forests adapt to the changing climate, while at the same time paying attention to the other needs for forests.

14.2   Multiple Forms of Knowledge and Expertise Required

The chapters in this book have included discussions on the feedback impacts between the natural biological world (forests) and social and technological processes (the technosystem), as well as the leakage impacts between regions. Moreover, it has become clear that, for research to derive results, it always needs to impose restrictions and assumptions, and analyse each phenomenon from some very particular perspective. Also, it is impossible to formulate alternative scenarios (counterfactuals) and evaluate their impacts with certainty. For example, in theory, we could compare the development of forest carbon sinks under two alternative scenarios involving wood harvesting levels. In one, the current level of annual wood harvesting in the EU is maintained, whilst in the other, the forest carbon sink is increasing due to a reduction in annual wood harvesting of 50% by 2050. What would be the impacts and differences between these two scenarios in terms of climate-change mitigation? The list of key impacts one would need to consider for this comparison is daunting—carbon sequestration in forests, the substitution impact of wood products, greenhouse gas (GHG) emissions technology development in non-wood sectors, carbon storage in wood products, the impacts of adaptation to the changing climate and forest disturbances, impacts on the forest carbon sink and wood harvesting in other regions (leakages), etc. Also, the analysis would need to be dynamic, making assumptions, for example, about what types of products forest products could substitute for in 2050 and how significant would their substitution impacts then be, or how much the changed climate at that time had increased the occurrence and impact of forest disturbances (forest fires, bark-beetle outbreaks, wind damage, etc.).

Moreover, it should be noted that the climate benefits from increased carbon sequestration in forests, gained by tailoring (adapting) forest management practices or by lowering wood-harvesting levels, may also be reinforced, counteracted or even offset by other concurrent changes. For example, by surface albedo, land-surface roughness, emissions of biogenic volatile organic compounds, transpiration and sensible heat flux (Luyssaert et al. 2018). Consequently, the tailoring of forest management could offset CO_2 emissions without halting the global temperature rise. However, it would probably be impossible for one study, or even a meta-study, to capture all these impacts. Even if it could, a number of restrictions and assumptions would need to be imposed, the realism of which would elicit many different views among scientists. Thus, currently *no absolute truth* of the impacts of different scenarios can be produced.

As Hulme et al. (2020) stated, "Technical and scientific knowledge is always partial, uncertain and often contradictory", but "that is not to say that such knowledge is not valuable… It is rather to say that to effectively deal with crises, multiple forms of knowledge and expertise are required, and political judgment is then necessary to sort, select and present it to public" (p. 4). Priebe et al. (2020) also explored a range of interacting obstacles that inhibited the increased use of forests as a climate-change mitigation tool. They state that "it is not a lack of knowledge or technical solutions that inhibits adapting measures to tackle climate change mitigation", but "current attempts to advise, guide, and implement sustainability suffer from an inability to examine and challenge prevailing values, habits, and ways of thinking" (p. 82).

Clearly, despite the complexity of the issue, the answer is not for us to raise up our hands and do nothing to mitigate climate change. Instead, scientific evidence should continue to be an important part of informing policy-makers, and the Intergovernmental Panel on Climate Change (IPCC), the Intergovernmental Science-Policy Platform on Biodiversity and Ecosystem Services (IPBES) and similar works are needed for assessing, synthesising and communicating this evidence for the making of policy (see Box 11.1).

In the scientific literature, there has been perhaps too much focus on the trade-offs between different actions and their impacts on mitigating climate change. Also, climate-change *mitigation* is often analysed separately from climate-change *adaptation* and *other societal aspirations*, such as reaching the United Nations' Sustainable Development Goals (SDGs). For example, Luyssaert et al. (2018) suggested that the primary role of forest management measures in Europe in the coming decades was not to protect the climate, but to adapt the forests to future climate in order to sustain the provisioning of wood and ecological, social and cultural services, while avoiding harmful (positive) climate feedbacks from fire, wind, pest and drought disturbances. Consequently, climate-change mitigation needs to be addressed simultaneously with other objectives, and the right balance of measures must be found that are politically possible to implement in the shortest time possible, because we are in a hurry to mitigate climate change.

Another issue that is important to bear in mind when interpreting scientific results is that the issue at hand can be more diverse and extensive than what the research may have considered. Let us illustrate this point with one example. In Chaps. 7 and 8, the role of wood as a substitute for fossil-based products was taken up. Related to this, a frequent suggestion from the research is that wood should not be used for short-lived products, such as energy and packaging, but instead for long-lived products that store carbon for a long period, such as wooden buildings. Also, the European Commission (2021) "leaked" Forest Strategy draft document recommended moving from short-lived wood products to long-lived ones. However, it is uncertain how workable this suggestion is in practice.

First, the world will not do without short-lived products, such as packaging, hygiene and textiles. They should also be made as low-carbon as possible. Short-lived products can also help to reduce CO_2 emissions. For example, food packaging helps to reduce food waste, and therefore also food production, which is associated with CO_2 emissions. Short-lived products may also be made from a different wood material than long-lived products—pulpwood, wood chips and production by-products (e.g. lignin) could be used. Logs are usually more suitable for producing long-lasting products, such as wooden buildings. Second, it might be possible that, in some cases, the net carbon mitigation impact of a short-lived forest-based product may be greater than for a long-lived product (Leskinen et al. 2018). This could also be possibly, for example, in the case when a country exports short-lived, wood-based textile fibres to China, where they help to replace synthetic, oil-based textiles in a manufacturing process that is also heavily coal based, versus using the wood fibre for some more long-lived product in the exporting country. For example, the EU27 exported 63% of its dissolving pulp in 2019, mainly to China and India. In these countries, dissolving pulp is used to replace synthetic (oil-based) fibres in the textile industry. Third, the climate-mitigation perspective is not the only important perspective; there are other possible environmental factors, such as plastics waste in the oceans or the quantity of materials used. Finally, the average service life of wood fibres in short-lived products could be substantially prolonged using recycling practices. For the reasons above, recommendations to use wood only in long-lasting products could be an oversimplification, and not necessarily optimal for climate mitigation.

Despite these complexities, it is self-evident that the forest-based sector can improve its performance in climate-change mitigation, for example, by reducing the use of fossil fuels in every part of the value chain, from harvest to the end-product market. Improvements in resource and production efficiency and circularity along the product chain can also decrease emissions and enhance biodiversity (e.g. less wood needs to be harvested, *ceteris paribus*).

Finally, when advancing a sustainable circular bioeconomy and tackling the grand challenges of the day, *discussion culture* is important. Unfortunately, we live in times in which some key politicians, parts of the traditional media and, especially, social media are enhancing societal polarisation. People seem to be taking evermore opposite and competing views, in which there is only black and white, with no shades of grey. This sometimes seems to rear its head in scientific discussion, or the science is used as a pretext for adopting clear positions and values (Pielke 2007; Hetemäki 2019). Opinions such as, "it is necessary to conserve all forests to act as carbon sinks" or "clear-cut harvesting is always a positive climate action", do not help us to reach urgently needed solutions. In this context, it is also important to monitor what type of perceptions the public gets from science and media, since perceptions shape opinions, media and voting, and therefore also political decisions.

14.3 Public Perceptions and Forest Bioeconomy

Public perception studies have shown that EU citizens appreciate forests mostly for the environmental services they provide; that is, as places for biodiversity, but also for their climate effects and the recreational opportunities they offer (Ranacher et al. 2020). However, Ranacher et al. (2020) also indicated that the potential role of the forest bioeconomy in climate mitigation is not well understood by the public. However, there are no clear research results that explain why this might be so. One guess is that this could be partly related to the fact that an increasing number of EU citizens live in urban areas, and they might be more inclined to appreciate the services that forests provide, rather than the products and welfare that is generated by the forest bioeconomy (Mauser 2021). In the EU in 2019, urban and peri-urban citizens accounted for a 75% share of the population, and therefore their views weight particularly strongly in public perceptions.[1]

For urban citizens, forests may have different meanings than for rural people and for those who live in and manage forests. Urban citizens may also be unaware of the benefits they derive from the forest-based sector. During an ordinary day, they may use or benefit from several wood-based products, such as buildings, furniture, food, packaging, clothing and energy, without realising that these are based on forests. Some of the benefits of the forest bioeconomy may be even more hidden; for example, in some EU countries with significant amounts of forests and forest industries (e.g. Austria, Estonia, Finland, Germany, Sweden), wood production and forest products help to generate income-, capital- and corporate-tax revenues, besides the more visible employment and income opportunities. These tax revenues can be used

---

[1] https://data.worldbank.org/indicator/SP.URB.TOTL.IN.ZS?end=2019&locations=EU& start=2019

to fund such things as social security, education and other societal infrastructure for the benefits of all citizens.[2]

Understandably, the forest-based sector and the benefits it generates may lie outside the urban bubbles in which the bulk of us live in the EU. Clearly, the forest-based sector has an interest and responsibility itself to communicate and inform the public of its sector and why it is important. In addition, to achieve a greater acceptance of the forest bioeconomy among citizens and policy-makers, it is important to provide facts about how the forest bioeconomy can be applied in order to respond to the more ambitious targets of climate-change mitigation and biodiversity conservation.

However, when EU or national policies are designed, it is also the responsibility of the European Commission, the European Parliament and national politicians to be informed and to be appreciative of *the many benefits*—not only some—that the EU forest-based sector provides for society. On the other hand, just as important is to acknowledge that the forest-based sector can also generate visible or hidden *disbenefits* for society, in terms of negative externalities, such as the loss or lack of biodiversity, the carbon sink, recreational opportunities and flood control. These disbenefits can be significant, especially if the forests are not managed and their bioproducts are not produced sustainably.

The European Green Deal (European Commission 2019) proposal did acknowledge the potential disbenefits of the forest sector and suggested important measures to tackle these. However, it failed to fully appreciate the potential benefits, such as a sustainable forest bioeconomy for climate mitigation and for achieving the SDGs (Palahí et al. 2020a, b). Emphasising only some of the benefits of forests is more likely to enhance the polarisation on this topic in society, which in turn could backfire by making it more difficult to further climate-mitigation and biodiversity objectives. In summary, it is essential that policy-makers have *a holistic approach* to the forest-based sector, and take into account all the many diverse impacts it can have, not just some.

14.4   The Role of EU Forests and the Forest-Based Sector

The European Green Deal (European Commission 2019) has set the overarching targets for EU policies in the coming years. At the heart of it is achieving climate neutrality by 2050, halting biodiversity loss and reaching the SDGs. In other words, paving the way to policies that will help the EU to live within the planetary boundaries. According to the messages coming from this book, how should the EU-forest-based sector help in this, and what types of policies could support this?

---

[2] For example, in Finland, the major forest-industry companies are the highest corporate tax-payers, the 10 largest of these alone paying €321 million in corporate taxes in 2019. For comparison, this is about the same amount that all the banks and insurance companies in the top 100 corporate tax-payers (17 companies) paid in 2019 (€335 million).

14.4.1 Role of the EU Forests

The urgency of mitigating climate change is key due to its potentially widespread and drastic impacts. Climate change also impacts all the other goals of the EU, such as biodiversity and the SDGs. The urgency itself makes things more difficult, especially in forests and the forest-based sector, which rely on slowly renewable—from decades to centuries—nature and wood. The urgency has also shifted political and public eyes to the land sector (agricultural land and forests) for help in reaching the climate-mitigation goals. There is an expectation that the speed at which we can reduce the root cause of climate change—burning fossil raw materials—is too slow for the set targets. Therefore, simultaneously increasing carbon storage and the forest sink in the coming decades is necessary, despite the fact that this could potentially become an excuse for some to continue to use fossil materials. Nevertheless, the reality seems to be that the EU will need larger land-based sinks in order to reach carbon neutrality by 2050 (IPCC 2019; Simon 2020).

As Chap. 2 explained, the EU27 forests account for 3.9% of the world's forests. Given this, it can have only a marginal direct impact at the global level via increased forest carbon sequestration. But every region has to contribute to climate mitigation. The aggregate impact of different regions counts, in the end, towards the global forest sink. Moreover, the indirect impacts of increasing EU forest carbon sequestration in the coming decades may be even more important. The EU is one key region in which the forest area, the annual volume growth of growing stock and the forest carbon stock have increased in the last decades. What is notable is that this happened at the same time as the EU27 wood production increased by 43%, from 1990 to 2019 (FAOSTAT 2021). The forest area has increased through natural forest expansion and the afforestation of low-productivity agricultural lands. Improved forest management practices and changing environmental conditions (e.g. nitrogen deposition and climate change) have increased the annual volume growth, carbon sequestration and storage of the EU forests. These have also been increasing because the annual wood harvesting has clearly been lower than the annual volume growth of the forests for a number of decades. This example of how to continue to increase forest growth and the carbon stock is important for other, less successful regions.

The book suggest various ways in which forest carbon sequestration can be increased in the future. Accordingly, the intensity of forest management and harvesting, and the severity of climate change and the associated increases in natural forest disturbances, will together determine the future development of carbon sequestration and storage in EU forests. Increasing the use of tailored adaptive forest management measures, such as the site−/region-specific cultivation of different tree species and genotypes (improved regeneration material), adjusting the frequency and intensity of thinnings and rotation lengths, using forest fertilisation and growing mixed forests, may still help to increase carbon sequestration and enhance forest resilience in the EU under the changing climate. Carbon sequestration may also be increased by increasing the forested area through natural forest expansion and the afforestation of low-productivity agricultural lands. However, forest carbon sink can also decrease due to an increase in natural forest disturbances. Overall, though, carbon sequestration and sinks in EU forests are likely to increase in the coming decades, as long as the annual wood harvesting and natural drain remains lower than the annual volume growth of the forests.

Forest conservation can play an important role in achieving carbon neutrality in the EU by 2050, but it is difficult to see how it could be the whole solution. Recent evidence from those regions that have not managed or harvested their forests for a long period of time, and that have suffered from serious disturbances, including forest fires, bark-beetle outbreaks and storms, for example, points to this conclusion (Högberg et al. 2021). Such regions occur e.g. in Australia, California, Canada, Russia and, in the EU, the Czech Republic, Portugal and Spain. As a result, forests have been destroyed and large amounts of GHGs have been emitted to the atmosphere. According to Camia et al. (2021), the wood harvested due to natural disturbances reached over 100 million m^3 (22.8% of the total removals) in 2018, in just 17 of the EU Member States that were surveyed. Moreover, old, unmanaged forests seem to sequester carbon less than young, managed forests (Gundersen et al. 2021). Thus, not managing forests and conserving them (which is clearly needed for many reasons) may also pose serious risks for climate-change mitigation. Between the extremes of conservation and deforestation, there are options for sustainably managing forests in ways that can retain them for generations as a source of a wide variety of ecosystem services.

In summary, it seems apparent that conserving the bulk of EU forests may not be an optimal climate-mitigation strategy (Nabuurs et al. 2017; EU 2018; IPCC 2019). The question is more about synergies and trade-offs between forest carbon sinks and forest management intensity in the short and medium terms. The land use, land-use change and forestry (LULUCF) regulation (EU 2018) allows member states to increase their forest utilisation for industrial and energy-production purposes in the future, if they can maintain or strengthen their long-term carbon sinks in forests and wood products. The EU wants to see a climate-neutral pathway to 2050 where the sinks of the LULUCF sector and all GHG emissions caused by humans are taken into account.

14.4.2   Role of the EU Forest-Based Bioeconomy

In general, the EU forest-based bioeconomy is responding to all sustainability challenges from the viewpoint of economic and environmental concerns and the societal transition towards sustainability (European Commission 2018, 2020). The starting point is that, by using more and more efficiently renewable materials, the increasing demand for non-renewable raw materials in the world can be curbed (International Resource Panel 2019), and this can lead to a more sustainable future, given that the biomass resources are used in sustainable way.

The updated EU Bioeconomy Strategy (European Commission 2018) highlights actions that will lead the way towards a sustainable and circular bioeconomy. They are: (1) strengthen and scale-up the bio-based sectors, unlock investments and markets; (2) deploy local bioeconomies rapidly across Europe; and (3) understand the ecological boundaries of the bioeconomy. In these ways, the Bioeconomy Strategy will maximise the contribution of the bioeconomy to the major EU policy priorities of sustainability, the creation of jobs, climate objectives, and the modernisation and strengthening of the EU industrial base. However, it is necessary to develop the bioeconomy in a way that lessens pressures on the environment, values and protects biodiversity and enhances all ecosystem services.

The role of, and necessity for, the forest bioeconomy in climate mitigation and the phasing out of fossil raw materials and products has been demonstrated in this book and in several studies (e.g. Hetemäki et al. 2017; Hurmekoski et al. 2018; IPCC 2019; Palahí et al. 2020a, b). Globally and in the EU, around 80–90% of the CO_2 emissions originate from the use of coal, oil and natural gas. Consequently, the core issue in the fight against climate change is the phasing out of fossil fuels and materials. If major efforts are not put into tackling these, they will remain a nuisance. In this context, it is difficult to see how climate mitigation can be possible without also using forest biomass to replace fossil-based raw materials and products. The forest bioeconomy is not going to be a sufficient way to solve the climate-change challenge on its own, but it is *a necessary part of it*.

In 2018 in the EU27, the GHG emissions were 3893 Mt. CO_2eq., of which 83.5% came from two sectors—energy production and industry (Eurostat data). In 2010–2016, EU forests helped to remove, on average every year, 10.4% of the total EU CO_2eq. emissions (Eurostat data). Including the impact of harvested-wood products, this figure was 11.3%, on average (Eurostat data). The EU is aiming to be climate-neutral by 2050—that is, a region with net-zero GHG emissions. Therefore, given the above figures, it is clear that the main priority should be to reduce emissions from fossil-based energy and industry to get them as close to zero as possible. Increasing EU forest removals will not reach this policy target. That is not to say that they are not important, or that the LULUCF regulation is needed to enhance this. Clearly, the EU has to do its share to increase the forest sink and removals, and in this way, show how it can be done. However, it is very important that the LULUCF regulation does not lead the debate and draw the focus of EU climate-change mitigation towards technical and relatively smaller issues, and away from the main issue itself, which is phasing out fossil fuels (Appiah et al. 2021, Berndes et al. 2018). This implies paying more attention to the need to develop new innovations in the forest bioeconomy, as well as improve the resource efficiency and circularity of current bioproducts.

In summary, it is essential that we use forest-based bioproducts for the same purposes we are currently using oil, coal and gas. However, given that it is unrealistic to phase out all fossil production by 2050, any remaining GHGs from these need to be balanced with an equivalent amount of carbon removal, for example by increasing the forest sink and through direct carbon capture and storage (CCS) technologies. Indeed, as Nabuurs et al. (2017) have argued, the EU can significantly increase the forest-based sector mitigation impact through forest removals and forest-product substitution impacts. These can be achieved by introducing new climate-smart forestry measures, and it is essential to seek to utilise this opportunity.

14.5 Combining Climate Mitigation with Other Goals

In the EU, climate mitigation is a top priority, but not the only one. The SDGs are also important priorities, and these include responsible consumption and production, sustainable cities and communities, and affordable and clean energy, among other things. According to the statistics reported in Chap. 1 (Box 1.1), employment in the forest bioeconomy in the EU28 was about 2.5 million, generating around €277 billion value added, in 2015. Moreover, wood-based construction, textiles and packaging are viewed as promising ways to make the EU's construction, clothing and packaging industries more sustainable.

Phasing out fossil-based industries will create a need to replace lost jobs (see Chap. 4, Box 4.3). Moreover, as the EU has emphasised, the Just Transition Mechanism is a key tool for ensuring that "the transition towards a climate-neutral economy happens in a fair way, leaving no one behind" (European Commission 2020). Even if the climate disaster looms with a 2, 3 or 4 °C temperature rise, it is possible that people could still reject the societal transition if they believe it to be unjust. As the former American Secretary of State, Henry Kissinger, has said *"No policy—no matter how ingenious—has any chance of success, if it is born in the minds of a few and carried in the hearts of none"*. The better the climate mitigation measures support other economic and societal goals and needs, the wider and stronger support they are likely to get amongst the citizens of the EU Member States. As a result, the mitigation measures could be adapted more promptly, and their implementation could be more efficient.

In summary, the circular forest bioeconomy may meet many diverse societal needs in the EU, along with its climate mitigation impact. Clearly, the success of meeting all these needs depends on how well the Member States are also able to impose the environmental sustainability of the forest bioeconomy. For this to happen, improving forest management and better adapting forests to the changing climate, increasing the resource efficiency of forest bioeconomy products, their circularity and new product innovations, as well as monitoring the bioeconomy's environmental sustainability are a must.

Key Messages

1. The forest-based sector[3] contributes to climate change mitigation via three channels—*forests are a carbon sink, forest-based products can substitute for fossil based-products, and these products can store carbon for up to centuries*. However, in order to produce climate benefits, the possible loss of carbon stock (sinks) in forests due to harvesting should be smaller than the increased

---

[3] The forest-based sector is here understood to include forests, forestry and forest-based products and energy.

GHG benefits of wood utilisation in the selected time frame. To achieve this mitigation objective in the future, forests also need to be resilient to changing climate. Therefore, *mitigation and adapting forests to climate change are married*, and both need to be advanced simultaneously.

2. The root cause of climate change and the scale of the impacts of different measures to mitigate it are important to keep in mind. Sometimes in the climate discussion, these self-evident facts seem to get lost, and small and large measures and impacts may get mixed. Globally and in the EU, around 80–90% of the CO_2 emissions originate from the use of coal, oil and natural gas. Consequently, *the core issue in the fight against climate change is the phasing out of fossil fuels*. It is essential to acknowledge that reaching this goal will not be possible without also using forest-based bioproducts to substitute for the current use of oil, coal and gas. In the EU forest-based context, this implies *paying more attention to the need to develop new innovations in the forest bioeconomy, improving the resource efficiency and circularity of current bioproducts, and imposing and monitoring the environmental sustainability of the bioeconomy*.

3. The optimal strategy to use forests and the forest-based sector to mitigate climate change, and to adapt them to the changing climate, requires *a holistic approach*. There is no single, optimal way for the forest-based sector to contribute to maximising mitigation and adaptation gains. Conserving forests only for carbon sequestration (storage, sinks) or using forests only for producing wood for forest bioproducts will not work. Both are needed for many different reasons, as the chapters in this book have explained. Moreover, *the optimal strategy needs to be tailored to the regional circumstances and characteristics. It may also need to be adjusted frequently over time as climate change proceeds*.

4. European forests belong mainly to the boreal and temperate forests. The lifecycles of the trees in these forests range from less than 100 years in managed forests to several hundred years in natural forests. The harvesting of trees to produce wood products takes place over a range of 20–100 years after one forest regeneration. Therefore, *when making decisions on forests, it is essential to keep in mind a time horizon of up to a century*. However, this is becoming increasingly difficult in our evermore rapidly changing world, in which continuous change is the norm. Also, the urgency of mitigating climate change and halting the loss of biodiversity call for rapid actions. This situation heightens the importance of the holistic approach and *the involvement of all science disciplines and societal perspectives* to plan sustainable actions regarding the entire forest-based sector. No single political party or interest group is likely to have the wisdom, and perhaps not always even the interest, to see the holistic picture.[4]

5. The risks of large-scale disturbances induced by weather extremes, such as droughts, storms and forest fires, have to be taken into account when appropri-

---

[4] American writer Upton Sinclair once stated: *"It is difficult to get a man to understand something when his salary depends upon his not understanding it."*

ate forest management measures are being defined over time to adapt to the changing climate. Carbon circulates between the atmosphere and forests either through natural cycles (e.g. decaying litter and organic matter and forest fires), or through harvesting and the use of wood for materials and energy. A viable forest sector enables the management of forests over large areas and adjustments to forest management measures, when needed. *The recovery of damaged wood or the reduction of the fuel load in forests is possible on a large scale, if there is a techno-system to enable both the harvesting and use of wood for the needs of society.*

6. The willingness of forest owners and society to adopt certain forest management measures depends on how they impact the other benefits generated by forests. The more synergies that can be found and the fewer trade-offs between them, the more likely and effectively they can be implemented. In short, *the effectiveness of the management measures needs to be assessed in their socio-economic context.*

7. The EU is not an island and its activities have impacts beyond its borders, for better or worse. It can serve as a good example to other regions of how ambitious climate and biodiversity goals can be achieved simultaneously. It can also demonstrate how the synergies can be maximised and the trade-offs minimised between a circular bioeconomy, climate-change mitigation and the maintenance of biodiversity. On the other hand, the EU climate mitigation and biodiversity policies can have negative *leakage impacts* on the climate-change mitigation and biodiversity in non-EU countries (Kallio et al. 2018; Dieter et al. 2020). Consequently, *the EU should assess the impacts of its policies in the global context, not only within its own borders.*

8. The world states are evermore interconnected, and therefore the problems they face tend to be increasingly global in nature, such as climate change, biodiversity loss, economic crises and pandemics. However, the consequences of such crises, and how they are solved, vary significantly, depending on regional features, such as national institutions and decision-making. How countries have been dealing with the COVID-19 pandemic is a good example of this. The nexus between forests, the bioeconomy and climate-change mitigation is no different. These things are linked by global challenges and opportunities, but their optimal implementation requires tailoring to regional and local circumstances—*one size does not fit all.* It is essential to acknowledge this when the EU is planning policies related to forests, the bioeconomy and climate-change mitigation for its Member States. *It can be argued that the stronger the EU is, the better it will succeed in coordinating common actions, but with Member State level tailoring and optimisation.*

9. It is important that science-based bodies like the IPCC and the IPBES make syntheses of the available knowledge and inform policy-making. However, the technical and scientific knowledge is always partial, uncertain and can even be contradictory. That is not to say that such knowledge is not valuable and needed. Rather, it points to the fact that to effectively deal with global-scale problems like climate change, multiple forms of knowledge and expertise are required. Moreover, how to best use the forest-based sector to mitigate and adapt to climate

change is not just a reducible engineering-type of problem, like how to get a man to the moon. ***Rather, it is a complex technological, economic, social, cultural and value problem***. "In the end the decisions are made by policymakers, and therefore, the decisions are political not scientific. In most societies, these decisions rest on democratic mandate, and so it should be" (Hetemäki 2019, p. 15).

10. The fundamental transformation of our society to carbon neutrality and sustainability is probably the greatest socio-political question we have faced since World War II. ***It has to be carried out in a way that people see it as just***. Otherwise, there is a danger that the whole process will be derailed and the transition will not happen. This is also true in the EU forest-based sector.

Box 14.1: Science Role in Informing Policy-Making

Lauri Hetemäki
European Forest Institute, Joensuu, Finland
Faculty of Agriculture and Forestry, University of Helsinki, Helsinki, Finland

Politicians, the media and the public, especially in the EU, are increasingly demanding evidence-based information to inform policy-making on complex issues such as climate change and biodiversity (Hetemäki 2019). According to Gluckman and Wilsdon (2016), "Scientific advice to governments has never been in greater demand; nor has it been more contested". Populist 'post-truth' politicians and social media warriors have questioned the legitimacy of science-based information. Ex-President Trump's questioning of the scientific evidence on climate change is one well-known example. But science skeptics can be found in Europe, as well, such as the senior British politician Michael Gove, who stated, during the Brexit referendum, that "people in this country have had enough of experts".

The matter is made more complex by the fact that there have occasionally been striking disparities between what the scientists' messages are (Hetemäki 2019). At the same time, there is an increasing amount of science information available. According to UNESCO (2015), almost 1.3 million scientific articles were published in 2014 alone, and there were 7.8 million full-time-equivalent researchers in 2013. Moreover, evidence-based policy-making has been the subject of much debate in the literature, particularly through critiques that question assumptions about the nature of the policy-making process, the validity of evidence, the skewing in favour of certain types of evidence, and the potentially undemocratic implications (Pielke 2007; Parkhurst 2017; Hetemäki 2019). One concern with evidence-based policy-making is that it does not recognise the contested nature of evidence itself—an area that has been the subject of a large body of research in the fields of the sociology of science and science and technology studies (Pielke 2007; Parkhurst 2017). These studies have drawn attention to the inevitably value-laden nature of scientific inquiry, and the choices that are made about what to

(continued)

Box 14.1 (continued)

research and how to undertake that research. Moreover, the choice of evidence can be value-laden and political in itself.

In December 2020, the Royal Swedish Academy of Agriculture and Forestry (KSLA) organised a seminar with the title 'Research results on forests are not always evidence'.[5] It may seem strange that an academic and science-based institution would choose to organise such a seminar, especially at a time when science information is so often being contested. However, as one might expect, the KSLA seminar did not question the importance of science knowledge in informing policy, as such, but rather it wanted to raise concerns about how evidence-based policy support can also be used incorrectly, and to advance specific interests and agendas. The seminar explained that research findings, or the conclusions that are drawn from the research, are not always well substantiated. The question was asked whether scientists and science publishers were seeking to gain power over policy-making, and if so, what did this mean for the democratic process? A key objective, therefore, is how to find a reasonable balance between more use of evidence-based information, whilst critically assessing new research results and keeping policy-making in the hands of politicians not scientists.

The KSLA seminar also raised questions about the role science publishers and the European Commission, in particular, drawing on the recent example of an article by Ceccherini et al. (2020), published in one of the most esteemed science journals—*Nature*. Using satellite data, the article reported an increase of 49% in the European harvested forest area alongside a biomass loss of 69% for the period 2016–2018 relative to 2011–2015. The article suggested that these increases reflected expanding wood markets, encouraged by the EU's bioeconomy policies. The article was written by the staff of the European Commission's Joint Research Centre—an institute responsible for providing evidence-based policy support for the EU. The article received an exceptional amount of media publicity all over Europe, as well as major policy attention, especially within the European Commission. *Nature* also published an Editorial (Nature 2020) based on the article, from which it drew the following conclusions and policy implications: "This is an important finding. It has implications for biodiversity and climate-change policies, and for the part forests play in nations' efforts to reach net-zero emissions... the increase in harvested forest area has been driven, in part, by demand for greener fuels, some of which are produced from wood biomass. This increase in biomass products can, in turn, be traced to the EU's bioeconomy strategy... Meeting renewable energy targets means burning Europe's harvest... [and] forest exploitation cannot continue at the current rate".

(continued)

---

[5] https://www.ksla.se/aktivitet/forskningsresultat-om-skog-ar-inte-alltid-evidens/

Box 14.1 (continued)

In a response to Ceccherini et al. (2020), also published in *Nature*, 33 scientists from 13 European countries provided evidence that threw into doubt the conclusions of the JRC study (Palahí et al. 2021). They demonstrated that the large reported harvest changes resulted from methodological errors. These errors related to satellite sensitivity having improved markedly over the period of the assessment, as well as to changes in the forests associated with natural disturbances—drought- and storm-related dieback and treefalls—that are often wrongly attributed to timber harvests. Palahí et al. (2021) stated that: "We argue that the reported changes reflect analytical artefacts, with (1) inconsistencies in the forest change time series, (2) misattribution of natural disturbances as harvests, and (3) lack of causality with the suggested bioeconomy policy frameworks". In addition to Palahí et al. (2021), some European organisations responsible for providing national forest statics voiced serious concerns that the Ceccherini et al. (2020) results were in conflict with official data on forest harvests (see, also, Wernick et al. 2021). Even after these concerns and pointing out the errors, the European Commission continued to refer to Ceccherini et al. (2020) as a basis for policy planning related to EU forests, such as biodiversity and climate policies.[6]

The seminar and example of an erroneous scientific paper, as well as similar lessons learnt from science-policy work (Pielke 2007; Parkhurst 2017; Hetemäki 2019), should raise awareness that we be critical, and understand the different interests that may lie behind producing science-based evidence and its publication and use. For example, different interest groups and non-governmental organisations are always fighting for the attention of policy-makers and the media, and they have great skills in searching for and selecting (i.e. cherry-picking) the scientific papers that support their specific agendas, using only those to lobby politicians and gain the media spotlight (Herajärvi 2021). Scientists may also have their own agendas, which may affect their selection of research topics and their scope, as well as fine-tuning the messages they deliver. In summary, not all science-based evidence, its publication and use, are equally neutral, robust and helpful - even if these are published in the most esteemed science journals.

The answer to the above challenges is not to do less science or use less science-policy information and dialogue, but to do it better. For this, several actions can be helpful. First, make science-policy work holistic and multidisciplinary. The problems we are facing are becoming evermore complex, such

(continued)

---

[6] For example, the Commissioner for Environment, Virginijus Sinkevičius, still referred to Ceccherini et al. (2020) results as important and valid in an interview by the main Swedish newspaper Dagens Nyheter, 14 June 2021. https://www.dn.se/debatt/tillsatt-en-skogsberedning-for-att-bromsa-polariseringen/

Box 14.1 (continued)

as the nexus between forests, biodiversity, climate change and the bioeconomy. However, science has become ever more specialised and focused on very specific and detailed questions, since this is how science advances. It has to set so-called system boundaries and exclude some factors from the analysis for it to be manageable. Given these drawbacks, there seems to be no other way to escape from this narrow and partial perspective than to produce holistic and multidisciplinary science-policy reviews, assessments and synthesis studies, and tailor these to a format that decision-makers can absorb.

To inform policy-making, there is a need for more cooperation among scientists to synthesise cross-disciplinary and multi-perspective science knowledge. It would also be desirable if this could be carried out by scientists with different values, and not only among those who are like-minded. This is what, for example, the IPCC, IPBES and UN International Resource Panel seek to do. They synthesise and contextualise multidisciplinary, global scientific knowledge for better informed decision-making. Such works should become the norm rather than the exception also at the national level. However, to be able to do such work, new funding is needed. Science-policy work is not research, as such, but rather the synthesis of existing research. Therefore, it will not generally be able to draw resources from research funding institutions. There is also a need to develop new funding mechanisms for science-policy work.

In summary, for better-informed science-policy-making, there is an increasing demand for greater cooperation in synthesising cross-disciplinary and multi-perspective science knowledge. Moreover, when policy-makers and the media use science information, it is important for them to be able to carefully assess the robustness of the information and find out if it has broad support in the science community.

Therefore, *it is important that the people who work in, and make a living from, forests and the forest-based sector are engaged and treated in a just way, so that they feel ownership of, and are willing to contribute effectively to, the transition*.

References

Appiah Mensah A, Petersson H, Berndes G, Egnell G, Ellison D, Lundblad M, Lundmark T, Lundström A, Stendahl J, Wikberg P-E (2021) On the role of forestry in climate change mitigation. EGU General Assembly 2021, EGU21–16472. https://doi.org/10.5194/egusphere-egu21-16472

Berndes G, Goldmann M, Johnsson F, Lindroth A, Wijkman A, Abt B, Bergh J, Cowie A, Kalliokoski T, Kurz W, Luyssaert S, Nabuurs GJ (2018) Forests and the climate: manage for maximum wood production or leave the forest as a carbon sink? Report by the Royal Swedish Academy of Agriculture and Forestry. https://www.ksla.se/wp-content/uploads/2018/12/KSLAT-6-2018-Forests-and-the-climate.pdf. Accessed 10 Jan 2021

Biber P, Felton A, Nieuwenhuis M, Lindbladh M, Black K, Bahýl J, Bingöl Ö, Borges JG et al (2020) Forest biodiversity, carbon sequestration, and wood production: modeling synergies and trade-offs for ten forest landscapes across Europe. Front Ecol Evol 8:e547696. https://doi.org/10.3389/fevo.2020.547696

Camia A, Giuntoli J, Jonsson R, Robert N, Cazzaniga NE, Jasinevičius G, Avitabile V, Grassi G, Barredo JI, Mubareka S (2021) The use of woody biomass for energy production in the EU. European Commission. JRC science for policy report. Publications Office of the European Union, Luxembourg. ISBN 978-92-76-27867-2. https://doi.org/10.2760/831621, JRC122719

Ceccherini G, Duveiller G, Grassi G et al (2020) Abrupt increase in harvested forest area over Europe after 2015. Nature 583:72–77. https://doi.org/10.1038/s41586-020-2438-y

Díaz-Yáñez O, Pukkala T, Packalen P, Lexer M, Peltola H (2020) Multi-objective forestry increases the production of ecosystem services. Forestry 94:1–9. https://doi.org/10.1093/forestry/cpaa041

Dieter M, Weimar H, Iost S, Englert H, Fischer R, Günter S, Morland C, Roering HW, Schier F, Seintsch B, Schweinle J, Zhunusov E (2020) Assessment of possible leakage effects of implementing EU COM proposals for the EU biodiversity strategy on forestry and forests in non-EU countries. Thünen working paper 159. https://doi.org/10.22004/ag.econ.307498

EU (2018) Regulation (EU) 2018/841 of the European Parliament and of the Council of 30 May 2018 on the inclusion of greenhouse gas emissions and removals from land use, land use change and forestry in the 2030 climate and energy framework, and amending Regulation (EU) No 525/2013 and Decision No 529/2013/EU. https://eur-lex.europa.eu/legal-content/EN/TXT/?uri=uriserv:OJ.L_.2018.156.01.0001.01.ENG. Accessed 10 Jan 2021

European Commission (2018) A sustainable bioeconomy for Europe – strengthening the connection between economy, society and the environment: updated bioeconomy strategy. https://knowledge4policy.ec.europa.eu/publication/sustainable-bioeconomy-europe-strengthening-connection-between-economy-society_en. Accessed 10 Jan 2021

European Commission (2019) European Green Deal. Striving to be the first climate-neutral continent. https://ec.europa.eu/info/strategy/priorities-2019-2024/european-green-deal_en. Accessed 10 Jan 2021

European Commission (2020) A strong social Europe for just transitions. https://ec.europa.eu/commission/presscorner/detail/en/qanda_20_20. Accessed 10 Jan 2021

European Commission (2021) New EU Forest Strategy post-2020: draft "leaked" 16 June 2021

FAOSTAT (2021) Forestry production and trade online statistics. http://www.fao.org/faostat/en/#data/FO. Accessed 10 Jan 2021

Gluckman P, Wilsdon J (2016) From paradox to principles: where next for scientific advice to governments? Palgrave Commun 2:16077. https://doi.org/10.1057/palcomms.2016.77

Gundersen P, Thybring EE, Nord-Larsen T et al (2021) Old-growth forest carbon sinks overestimated. Nature 591:E21–E23. https://doi.org/10.1038/s41586-021-03266-z

Herajärvi H (2021) Science as a decision-support tool in forest policies. Silva Fenn 55(2). https://doi.org/10.14214/sf.10566

Hetemäki L (2019) The role of science in forest policy – experiences by EFI. For Policy Econ 105:10–16. https://doi.org/10.1016/j.forpol.2019.05.014

Hetemäki L, Hanewinkel M, Muys B, Ollikainen M, Palahí M, Trasobares A (2017) Leading the way to a European circular bioeconomy strategy. From science to policy 5. European Forest Institute. https://doi.org/10.36333/fs05

Högberg, P, Ceder, LA, Astrup, R, Binkley, D, Dalsgaard, L, Egnell, G, Filipchuk, A, Genet, H, et al (2021) Sustainable boreal forest management – challenges and opportunities for climate change mitigation. Swedish Forest Agency, Report 2021/11. https://www.skogsstyrelsen.se/globalassets/om-oss/rapporter/rapporter-2021202020192018/rapport-2021-11-sustainable-boreal-forest-management-challenges-and-opportunities-for-climate-change-mitigation-002. pdf. Accessed 27 Jan 2022

Hulme M, Lidskog R, White JM, Standring A (2020) Social scientific knowledge in times of crisis: what climate change can learn from coronavirus (and vice versa). Wiley Interdiscip Rev Clim Chang 11(4). https://doi.org/10.1002/wcc.656

Hurmekoski E, Jonsson R, Korhonen J, Jänis J, Mäkinen M, Leskinen P, Hetemäki L (2018) Diversification of the forest industries: role of new wood-based products. Can J For Res 48(12):1417–1432. https://doi.org/10.1139/cjfr-2018-0116

Intergovernmental Panel on Climate Change (IPCC) (2019) Climate Change and Land: an IPCC special report on climate change, desertification, land degradation, sustainable land management, food security, and greenhouse gas fluxes in terrestrial ecosystems. https://www.ipcc.ch/site/assets/uploads/sites/4/2021/02/210202-IPCCJ7230-SRCCL-Complete-BOOK-HRES.pdf. Accessed 10 Jan 2021

International Resource Panel (IRP) (2019) Global resources outlook 2019: natural resources for the future we want. UN Environment Programme and IRP https://wedocsuneporg/handle/2050011822/27517. Accessed 10 Jan 2021

Kallio M, Solberg B, Käär L, Päivinen R (2018) Economic impacts of setting reference levels for the forest carbon sinks in the EU on the European forest sector. Forest Policy Econ 92. https://doi.org/10.1016/j.forpol.2018.04.010

Krumm F, Schuck A, Rigling A (eds) (2020) How to balance forestry and biodiversity conservation. A view across Europe. European Forest Institute, Joensuu. https://doi.org/10.16904/envidat.196

Leskinen P, Cardellini G, González-García S, Hurmekoski E, Sathre R, Seppälä J, Smyth C, Stern T, Verkerk, H. (2018). Substitution effects of wood-based products in climate change mitigation. From science to policy 7. European Forest Institute. https://doi.org/10.36333/fs07

Luyssaert S, Marie G, Valade A, Chen Y-Y, Djomo SN, Ryder J, Otto J, Naudts K, Lanso AS, Ghattas J, McGrath MJ (2018) Trade-offs in using European forests to meet climate objectives. Nature 562:259–262. https://doi.org/10.1038/s41586-018-0577-1

Mauser H (ed) (2021) Key questions on forests in the EU. Knowledge to action 4. European Forest Institute, Joensuu. https://doi.org/10.36333/k2a04

Nabuurs G-J, Delacote P, Ellison D, Hanewinkel M, Hetemäki L, Lindner M (2017) By 2050 the mitigation effects of EU forests could nearly double through climate smart forestry. Forests 8:484. https://doi.org/10.3390/f8120484

Nature (2020) How Europe can fix its forests data gap. Editorial, 1 July 2020. https://www.nature.com/articles/d41586-020-01848-x. Accessed 10 Jan 2021

Palahí M, Hetemäki L, Potocnik J (2020a) Bioeconomy: the missing link to connect the dots in the EU Green Deal. EURACTIV press release, 20 March 2020. European Forest Institute, Joensuu. https://pr.euractiv.com/pr/bioeconomy-missing-link-connect-dots-eu-green-deal-202385. Accessed 10 Jan 2021

Palahí M, Pantsar M, Costanza R, Kubiszewski I, Potočnik J, Stuchtey M, Nasi R, Lovins H, Giovannini E, Fioramonti L, Dixson-Declève S, McGlade J, Pickett K, Wilkinson R, Holmgren J, Trebeck K, Wallis S, Ramage M, Berndes G, Akinnifesi FK, Ragnarsdóttir KV, Muys B, Safonov G, Nobre AD, Nobre C, Ibañez D, Wijkman A, Snape J, Bas L (2020b) Investing in nature as the true engine of our economy: a 10-point action plan for a circular bioeconomy of wellbeing. Knowledge to action 02. European Forest Institute, Joensuu. https://doi.org/10.36333/k2a02

Palahí M, Valbuena R et al (2021) Concerns about reported harvests in European forests. Nature 592:E15–E17. https://doi.org/10.1038/s41586-021-03292-x

Parkhurst J (2017) The politics of evidence: from evidence-based policy to the good governance of evidence. Routledge studies in governance and public policy. Routledge, Abingdon. ISBN 9781138939400. https://eprints.lse.ac.uk/68604/1/Parkhurst_The%20Politics%20of%20Evidence.pdf. Accessed 10 Jan 2021

Pielke RA (2007) The honest broker: making sense of science in policy and politics. Cambridge University Press, Cambridge

Priebe J, Mårald E, Nordin A (2020) Narrow pasts and futures: how frames of sustainability transformation limit societal change. J Environ Stud Sci. https://doi.org/10.1007/s13412-020-00636-3

Ranacher L, Sedmik A, Schwarzbauer P (2020) Public perceptions of forestry and the forest-based bioeconomy in the European Union. Knowledge to action 03. European Forest Institute, Joensuu. https://doi.org/10.36333/k2a03

Simon F (2020) Commission under fire for including 'carbon sinks' into EU climate goals. EURACTIVE 18 Sept 2020. https://www.euractiv.com/section/climate-environment/news/commission-under-fire-for-including-carbon-sinks-into-eu-climate-goals/. Accessed 10 Jan 2021

UNESCO (2015) UNESCO science report: towards 2030. https://unesdoc.unesco.org/ark:/48223/pf0000235406/PDF/235406eng.pdf.multi. Accessed 15 Feb 2021

Wernick IK, Ciais P, Fridman J et al (2021) Quantifying forest change in the European Union. Nature 592:E13–E14. https://doi.org/10.1038/s41586-021-03293-w

Epilogue

Lauri Hetemäki

A forest is the opposite of the cake in the adage, which you either eat or have. In order to have your forest cake, you must eat it; if you eat it properly, you get even more of it.

Egon Glesinger said this in 1949, but has it been evidenced in the last few decades (Glesinger 1949)?[1] Unfortunately, the world's forest area has been declining—for example, from 1990 to 2020 by 177 million ha—which is a little more than the total forest area of the EU27 (159 million ha) (FAOSTAT 2021). This deforestation has been taking place mainly in Africa and South America. But there are positive examples as well—with the help of appropriate forest management (among other factors), the EU27 has in the last three decades increased its forest area by 14 million ha (10%), equivalent to almost the combined total land area of Austria, Belgium and the Netherlands. The volume of wood in forests—the growing stock—has increased even more rapidly, from about 19 to 27 billion m^3, or 43% (Eurostat 2020). Yet, in the last three decades, the EU27 has also used 13 billion m^3 of roundwood for forest products and energy (FAOSTAT 2021). Thus, it seems that Glesinger's statement holds true: the EU27 has both eaten the forest cake substantially, but at same time has also made it much bigger.

But neither life, welfare nor the bioeconomy is primarily about quantity, but rather about quality. Simple forest and roundwood statistics hide other factors that we value and view as important in our forests, such as biodiversity, forest carbon sinks, recreation, culture—in short, all the forest ecosystem services. Although in the EU, the carbon stock in forests has been increasing over the past three decades, climate policy targets require this to be increased even more in the future. Also, there is significant agreement that biodiversity in the EU forests needs to be further enhanced.

How to handle all the different and increasing demands for forests in a balanced and sustainable way will remain a key policy and practical forest management issue

---

[1] Glesinger, E. (1949) The Coming Age of Wood. Simon and Schuster, New York. http://www.archive.org/details/comingageofwood00gles

© The Author(s) 2022
L. Hetemäki et al. (eds.), *Forest Bioeconomy and Climate Change*, Managing Forest Ecosystems 42, https://doi.org/10.1007/978-3-030-99206-4

in the decades to come. In this context, this book has argued for the need for a holistic approach, rather than sometimes-voiced partial actions and overly simplified instructions. Moreover, we have explained that the climate-smart forestry approach, as outlined in this book, tailored to regional circumstances, provides a useful way to move forward. This will require abandoning some conventional thinking, such as seeing the bioeconomy and climate mitigation or the bioeconomy and biodiversity, as separate and always in opposition to each other. All of these are needed, and with clever management, they can be mutually supporting. *Properly cared for, forests are great all-rounders—they produce climate benefits, biodiversity and recreation, food, as well as one of the most versatile and useful raw materials on earth.*

References

Eurostat (2020) Forests, forestry and logging. Statistics explained. https://ec.europa.eu/eurostat/statistics-explained/pdfscache/52476.pdf. Accessed 10 Jan 2021
FAOSTAT (2021) Forestry production and trade online statistics. http://www.fao.org/faostat/en/#data/FO. Accessed 10 Jan 2021

Glossary

Accounting of GHG emissions Further calculations on reported GHG data, for example, by comparing annual emissions and sinks to a baseline. This facilitates tracking progress towards set mitigation targets.

Albedo Reflection of sunlight from surfaces. Darker colours reflect less sunlight and absorb more energy, hence causing local warming.

Bioeconomy Many different definitions exists. One useful definition is from the Global Bioeconomy Summit (GBS) 2015: *"bioeconomy as the knowledge-based production and utilization of biological resources, innovative biological processes and principles to sustainably provide goods and services across all economic sectors"*. The bioeconomy therefore encompasses the traditional bioeconomy sectors, such as forestry, paper and wood products, as well as emerging new industries, such as textiles, chemicals, new packaging and building products, biopharma, and also the services related to those products (research and development, education, sales, marketing, extension, consulting, corporate governance, etc.), and forest services (recreation, hunting, tourism, carbon storage, biodiversity, etc.).

Biogenic carbon emissions Carbon dioxide emissions originating from biological sources such as organic soils or tree biomass. Albeit the biogenic carbon dioxide molecules are identical to those originating from fossil sedimentations, it can be useful to track them separately for analytical purposes. This is because fossil emissions lead to a permanent increase in the CO_2 concentration of the atmosphere, while biogenic carbon is eventually absorbed back to biomass when trees grow, if a harvest site is not permanently deforested.

Carbon balance Balance between carbon sequestration and emissions at a given moment or over a given period.

Carbon dioxide (CO_2) Green house gas. Human activities since the beginning of the Industrial Revolution (around 1750) have increased the atmospheric concentration of carbon dioxide by almost 50%, from 280 ppm in 1750 to 419 ppm in 2021. The last time the atmospheric concentration of carbon dioxide was this high was over 3 million years ago.

© The Author(s) 2022
L. Hetemäki et al. (eds.), *Forest Bioeconomy and Climate Change*, Managing
Forest Ecosystems 42, https://doi.org/10.1007/978-3-030-99206-4

Carbon emission Carbon released in living processes and the decomposition of organic matter with carbon.

Carbon leakage Carbon leakage is a form of rebound effect, typically referring to a shift of emission intensive production from one region to another, for example, to escape restrictive regulation. It can occur in a geographic (international carbon leakage), product substitution (intersectoral carbon leakage) or temporal (intertemporal carbon leakage) dimension.

Carbon sequestration Removal and storage of carbon from the atmosphere in carbon sinks in forests through photosynthesis.

Carbon sink Forest ecosystem taking up more carbon than it emits.

Carbon source Forest ecosystem taking up less carbon than it emits.

Carbon stock (storage) Amount of carbon in the ecosystem or its compartments (e.g. above- or below-ground, trees, soil etc.) at a given moment or over a given period.

Carbon uptake Carbon fixed in photosynthesis and converted to different tissues in growth of biomass.

Cascade use of wood Recovery and ensuing reuse, recycling, or incineration of a forest-based product to extend the lifecycle of biomass in the techno-system.

Circular bioeconomy A circular bioeconomy builds on the mutual efforts of the circular economy and bioeconomy concepts, which in many ways are interlinked. The European Environment Agency (EEA) has indicated that implementing the concepts of a bioeconomy and circular economy together as a systemic joint approach would improve resource efficiency and help reduce environmental pressures (EEA 2018). We further suggest that these two concepts, which are often considered separately, could create marked synergies when applied as a hybrid approach, making simultaneous use of both, as is the concept of the circular bioeconomy.

Climate-smart forestry A holistic approach to how forests and the forest-based sector can contribute to climate-change mitigation that considers the need to adapt to climate change, while taking into account the specific regional setting. It builds upon three main objectives: (1) reducing and/or removing GHG emissions; (2) adapting and building forest resilience to climate change; and (3) sustainably increasing forest productivity and incomes.

Climate change adaptation Climate change adaptation is the process of adjusting to current or expected climate change and its effects. For humans, adaptation aims to moderate or avoid harm or minimize risks of harm, and exploit opportunities; for natural systems, humans may intervene to help adjustment.

Climate change mitigation Climate change mitigation consists of actions to limit global warming and its related effects. This involves reductions in human emissions of greenhouse gases (GHGs) as well as activities that reduce their concentration in the atmosphere.

Counterfactual scenario Depicts the state of a system, if a scenario would not have been realised. For example, what the net carbon emissions of the sector would have been, if the level of harvest had been different.

Creative destruction The term was coined by Joseph Schumpeter in the 1940s. It refers to the continuous process of entire industries rising and falling as a result of established goods and services getting substituted by innovations and changes in the operating environment.

Decomposition Carbon lost in the heterotrophic respiration.

Demand An economic principle depicting consumers' willingness to purchase goods and services. In practice it means the amount or value of a good consumed.

Derived demand For non-consumer products, the demand can be modelled by assuming that the demand for a forest-based product is a function of the same factors that affect the demand for the final uses of the product.

Displacement factor A measure for the amount of fossil GHG emissions avoided per one unit of HWPs consumed in place of a specific alternative product.

Diversification Diversification refers to markets becoming more heterogeneous. In the context of forest-based products, it refers to the declining trend in some of the large volume product groups and the simultaneous emergence of new product groups.

Ecosystem services The Millennium Ecosystem Assessment, a United Nations report describing the condition and trends of the world's ecosystems, categorizes ecosystem services as: (1) Provisioning Services such as food, clean water, fuel, timber, and other goods; (2) Regulating Services such as climate, water, and disease regulation as well as pollination; (3) Supporting Services such as soil formation and nutrient cycling; and (4) Cultural Services such as educational, aesthetic, and cultural heritage values, recreation, and tourism.

ESM Earth system models simulate all relevant aspects of the Earth system. They include physical, chemical and biological processes, and therefore reach beyond their predecessors, the global climate models (GCM), which just represented the physical atmospheric and oceanic processes.

Forest sector A term used to describe forestry and forest industries. The Forest Sector Outlook Studies (UNECE-FAO) have defined the forest sector as to cover both, forest resources and the production, trade and consumption of forest products and services. Forest products include all the primary wood products manufactured in the forest processing sector (sawnwood, wood-based panels, paper and paperboard) and the main inputs or partly processed products used in the sector (roundwood, wood pulp, wood residues and recovered paper). Secondary or value-added forest products (such as wooden doors, window frames and furniture) are not covered. *Forest-based sector:* Is a more recent and more extensive concept than the "forest sector". In addition to forest sector, it includes the whole value chains from basically all industries (or industry sectors) that use wood as a main raw-material, such as forest based bioenergy, biochemicals, biotextiles, construction, packages, hygiene products, etc.

Forest reference level A fixed target level for the forest carbon sink. It is used, for example, in the accounting of emissions and sinks in the EU LULUCF regulation.

Functional unit A reference to which the inputs (raw materials and land use) and outputs (emissions) of a good or service are calculated, such as a square meter of a multi-storey building with a specific design and properties.

GCM Global Climate Model

GDD Growing Degree Day

Green house gas A greenhouse gas (GHG or GhG) is a gas that absorbs and emits radiant energy within the thermal infrared range, causing the greenhouse effect. The primary greenhouse gases in Earth's atmosphere are water vapor (H_2O), carbon dioxide (CO_2), methane (CH_4), nitrous oxide (N_2O), and ozone (O_3).

Gross primary production Carbon fixed in photosynthesis.

Growing stock Growing stock is the volume of all living trees and excludes smaller branches, twigs, foliage and roots. It is measured in cubic metres (m^3) over bark and includes trees of more than a given size (in terms of diameter) at breast height (Eurostat, "Forestry in the EU and the world. A Statistical Portrait" (2011)).

Harvested Wood Products (HWPs) The term Harvested Wood Product is commonly used in technical literature and policy documents, for example, by the IPCC and UNFCCC when referring to changes in the carbon storage of wood-based products in GHG reporting. It can be used as a synonym for forest-based products.

Humus Dead organic matter in soil having a non-identifiable origin.

HWP pool All forest-based products present in the techno sphere.

IPBES The Intergovernmental Science-Policy Platform on Biodiversity and Ecosystem Services (IPBES) is an independent intergovernmental body established by States to strengthen the science-policy interface for biodiversity and ecosystem services for the conservation and sustainable use of biodiversity, long-term human well-being and sustainable development. It was established in 2012 by 94 Governments. It is not a United Nations body. However, at the request of the IPBES Plenary and with the authorization of the UNEP Governing Council in 2013, the United Nations Environment Programme (UNEP) provides secretariat services to IPBES.

Industrial prefabrication Off-site manufacturing of elements and components, which allows combining several work phases in a single off-site location in standardised, conveyor belt type of conditions

Inferior good The demand for an inferior good decreases when income increases, and increases when income decreases

IPCC International Panel on Climate Change (IPCC) was created by the United Nations Environment Programme (UN Environment) and the World Meteorological Organization (WMO) in 1988, the IPCC has 195 Member countries. The IPCC prepares comprehensive Assessment Reports about the state of scientific, technical and socio-economic knowledge on climate change, its impacts and future risks, and options for reducing the rate at which climate change is taking place.

Life cycle assessment (LCA) The practice of assessing the environmental impacts of products or processes using standardised indicators.

Litter Dead organic matter in soil having an identifiable origin (e.g. needle litter).

LULUCF sector Land use, land use change and forestry sector. It is one of the sectors in the classification of human activities in GHG reporting, comprising forest land, cropland, grassland, wetlands, settlements and other land.

Market Markets consist of suppliers and consumers of a given product. The prices and quantities of products are in equilibrium as a consequence of intersecting supply and demand curves.

Net Annual Increment (NAI) Average annual volume over the given reference period of gross increment less that of natural losses on all trees to a minimum diameter of 0 cm (d.b.h.). Source: http://www.unece.org/forests/fra/definit. html#Net%20annual.

Net ecosystem exchange (NEE) A measure of the net exchange of CO_2 between an ecosystem and the atmosphere.

Net ecosystem production Heterotrophic respiration deducted from net primary production

Net primary production Carbon lost in plant respiration deducted from gross primary production.

New forest-based products Products in the introduction, growth or renewal phase of the product life cycle. The demand for new forest-based products is mainly determined by factors unrelated to economic activity.

No-regret pathway A scenario that depicts measures for achieving desired outcomes with minimum trade-offs between different objectives.

Normal good The demand for a normal good increases when income increases, and decreases when income decreases

Operating environment An umbrella term for the external factors influencing the markets of a given industry, often categorised into political, economic, social, technological, environmental and legal factors.

Outlook study Outcome of a foresight exercise that applies various futures research methods.

Price elasticity A unit change in demand as a result of a unit change in the price of a given good. It can be perceived as the slope of the demand curve. Elasticities can also be determined for various demand shifters such as income.

Product lifecycle The entire lifespan of a product from introduction to saturation to decline in demand. Note the difference between the lifespan of a single product from production to disposal (used in life cycle assessment) and the lifespan of a product as a part of an industry (used in economics).

RCP Representative Concentration Pathway is a greenhouse gas concentration trajectory adopted by the IPCC.

Rebound effect An additional unit of a product consumed does not lead to a unit reduction in other consumption, due to indirect impacts to e.g. available income.

Reporting of GHG emissions Calculation and publication of the emissions and sinks of all sectors, following jointly agreed technical guidance. Reporting of GHG emissions: See Chap. 8.

Resilience In science, there are many dimensions of resilience concept. For example, ecological resilience is the capacity of an ecosystem to respond to a perturbation or disturbance by resisting damage and recovering quickly. Such perturbations and disturbances can include events such as fires, flooding, windstorms, insect population explosions, and human activities such as deforestation, pesticide sprayed in forests, and the introduction of exotic plant or animal species. Climate resilience is the ability of systems to recover from climate change.

Roundwood Wood in its natural state as felled, with or without bark. It may be round, split, roughly squared or in other forms. Roundwood can be used for industrial purposes, either in its round form (e.g. as transmission poles or piling) or as raw material to be processed into industrial products such as sawn wood, panel products or pulp; or it can be used for energy purposes (fuel wood).

Structural change A significant and permanent change in the structure of an industry, such as a change in the shares of sub-industries from the total production. If caused by substitution, a structural change can be detected by a statistical test for income elasticity.

Substitute product See Chap. 4.2.2.1 glossary

Substitution impact Avoided fossil emissions, when wood products are used in place of more fossil emission intensive products. Note that this does not signify the climate change mitigation potential of forest-based products, but is an important determinant for it.

Substitution An increase in demand for one good in place of another good. A perfect substitute provides interchangeable value or service, either in terms of economic utility or technical function.

Sustainability The most often quoted definition comes from the UN World Commission on Environment and Development: "sustainable development is development that meets the needs of the present without compromising the ability of future generations to meet their own needs." In the charter for the UCLA Sustainability Committee, sustainability is defined as: "the integration of environmental health, social equity and economic vitality in order to create thriving, healthy, diverse and resilient communities for this generation and generations to come. The practice of sustainability recognizes how these issues are interconnected and requires a systems approach and an acknowledgement of complexity."

Sustainable Development Goals (SDG) The Sustainable Development Goals (SDGs) or Global Goals are a collection of 17 interlinked global goals designed to be a blueprint to achieve a better and more sustainable future for all. The SDGs were set up in 2015 by the United Nations General Assembly and are intended to be achieved by the year 2030. They are included in a UN Resolution called the 2030 Agenda or what is colloquially known as Agenda 2030.

Techno sphere (techno-system) A generic concept for differentiating the system boundaries between ecosystems and man-made systems. It includes all products and processes outside ecosystems and may refer to the entire lifespan of forest-based products after harvesting.

Trend forecast A simplistic depiction of the future direction of a time series, if it would develop along the lines of a selected historical period. It should not necessarily be regarded as a prediction, but a helpful baseline, against which possible deviations from the trend can be assessed.

UNFCCC United Nations Framework Convention on Climate Change is a UN body that facilitates the process under which global climate agreements such as the Paris Agreement are negotiated among governments.

UNEP United Nations Environment Programme is responsible for coordinating the UN's environmental activities and assisting developing countries in implementing environmentally sound policies and practices.

Value chain A value chain describes the full range of activities which are required to bring a product or service from conception, through the different phases of production (involving a combination of physical trans-formation and the input of various producer services), delivery to final consumer, and final disposal after use.

WMO World Meteorological Organization is a specialized agency of the United Nations and responsible for promoting international cooperation on atmospheric science, climatology, hydrology and geophysics.

Printed in the United States
by Baker & Taylor Publisher Services